Royal Stars
OF THE STATES

50 MAJESTIC QUILTS WITH COMPLETE INSTRUCTIONS

HOUSE of
WHITE
BIRCHES

PUBLISHERS
SINCE 1947

Royal Stars of the States

Editor: Sandra L. Hatch
Designs: Eula Mae Long
Diagrams: Betty Boyink
Quiltmaker: Dolores Yoder
Copy Editor: Cathy Reef
Assistant to Editorial Director: Jeanne Stauffer
Editorial Director: Vivian Rothe

Photography: Klaus Rothe, Vicki Macy

Production Manager: Vicki Macy
Creative Coordinator: Shaun Venish
Technical Artist: Connie Rand
Production Coordinator: Sandra Ridgway Beres
Production Assistants: Patricia Elwell, Cheryl Lynch, Matt Martin, Andrew Staub

Publishers: Carl H. Muselman, Arthur K. Muselman
Chief Executive Officer: John Robinson
Marketing Director: Scott Moss

Printed in the United States of America
First Printing: 1988; Second Printing: 1995
Library of Congress Number: 95-78329
ISBN: 1-882138-15-5

Royal Stars of the States first appeared in serial form in *Quilt World* magazine, published by House of White Birches.

Every effort has been made to ensure the accuracy and completeness of the instructions in this book. However, we cannot be responsible for human error or for the results when using materials other than those specified in the instructions, or for variations in individual work.

Cover quilt: *Royal Star of Virginia*, pattern begins on page 170.

Royal Stars

	PHOTO	INSTRUCTIONS		PHOTO	INSTRUCTIONS
Alabama	7	62	New Hampshire	35	131
Alaska	8	65	New Jersey	36	133
Arizona	9	67	New Mexico	37	136
Arkansas	10	69	New York	38	138
California	11	72	North Carolina	39	140
Colorado	12	75	North Dakota	40	142
Connecticut	13	78	Ohio	41	144
Delaware	14	81	Oklahoma	42	146
Florida	15	83	Oregon	43	148
Georgia	16	87	Pennsylvania	44	152
Hawaii	17	88	Rhode Island	45	154
Idaho	18	90	South Carolina	46	156
Illinois	19	93	South Dakota	47	158
Indiana	20	95	Tennessee	48	160
Iowa	21	98	Texas	49	164
Kansas	22	102	Utah	50	166
Kentucky	23	104	Vermont	51	168
Louisiana	24	107	Virginia	52	170
Maine	25	110	Washington	53	173
Maryland	26	112	West Virginia	54	177
Massachusetts	27	114	Wisconsin	55	178
Michigan	28	115	Wyoming	56	180
Minnesota	29	117	Border 1		183
Mississippi	30	118	Border 2		185
Missouri	31	121	Border 3		188
Montana	32	123	Border 4		189
Nebraska	33	126	Border 5		191
Nevada	34	129			

OF THE STATES

50 MAJESTIC QUILTS WITH COMPLETE INSTRUCTIONS

A Quilted Tribute to All 50 States

The glorious quilts in this star-studded collection were designed by Eula Mae Long, a quilter who was moved to create the Royal Stars of the States after she was asked if she had a pattern for an Oregon Star. Eula Mae submitted her star quilt patterns to *Quilt World* magazine in 1982 for publication so that other quilters could also enjoy creating the beautiful designs. The patterns for the 50 star quilts were published on a regular basis in *Quilt World* during the following five years. Throughout that time, many readers wrote to House of White Birches requesting that all of the Royal Star patterns be printed as a collection in one book. We were pleased to grant their request, first with a soft-cover edition published in 1988. For the past seven years the book has continued to be very popular with quilters, and we have responded with this new, revised and expanded hardcover edition.

Several quilters have actually completed all 50 quilts from the *Quilt World* series! One of those quilters is Dolores Yoder, and it is photographs of her Royal Stars quilt collection that we share with you in this book. You'll enjoy reading more about this fascinating Texas quiltmaker on page 6.

As noted in the General Instructions on pages 57–58, the patterns and instructions for the Royal Star quilts are intended for those quilters who have mastered the basics of quiltmaking and are looking for a challenge. The directions do not include detailed how-to's or hints for fast-piecing techniques, which can be used for many of the patterns. Instead, the patterns provide a great opportunity to apply the latest techniques or skills you might have learned at your recent quilt class. They also give the experienced quilter a wonderful opportunity to use her creativity in her choice of colors and in her selection of a border pattern. Dolores Yoder's color choices are shown in this book, but don't let that limit your personal preferences!

In response to the requests from a number of quilters, we are including in this new edition of *Royal Stars of the States* five versatile border patterns that can be used to frame your quilt. Eula Mae Long's Royal Star quilt designs serialized in *Quilt World* magazine did not include any border patterns. The quilts pictured in this book show the border patterns Dolores Yoder selected to frame her Royal Star quilts. You can choose a border to complement your quilt from any of the five patterns beginning on page 182, or you can choose a border pattern of your own.

Star quilts are great crowd pleasers and ribbon winners. The serious quilter will find many designs in this collection of star patterns that are perfect for that prizewinning quilt, charity quilt auction or club raffle. You'll have hours of enjoyment studying these splendid designs, choosing just the right one to stitch.

When you have completed a quilt for your own home state, you'll want to try some of the designs for the other states. You may want to share your handiwork with others through your local quilt shows and community festivals. You'll find everyone agrees that a beautiful Royal Star quilt is a much appreciated way to honor your home state. We have also heard from quilters in foreign countries who have made quilts from these patterns. It's exciting to know that our American quilt patterns are appreciated throughout the world!

May you enjoy many pleasurable hours creating one of these lovely works of art!

Eula Mae Long

Designer of the Royal Stars of the States Quilts

Many of us leave behind evidence of our existence on this earth in unusual ways. Some of us have children, make quilts, write books or paint. Eula Mae Long's legacy will be her Royal Stars of the States quilt designs. These designs are being made by quilters all over the country. Because these quilts will most surely survive into the next century, Eula Mae's place in quilt history has been assured.

Eula Mae Long was born in Prescott, Iowa, in 1925. She was the only girl in a house of six boys. She began sewing at the age of 9 or 10 when she began a Sunday afternoon ritual. Eula Mae grew to hate those afternoon sessions because her mother insisted her stitches be small and consistent. Eula Mae's aversion to hand piecing lingers today; therefore, she prefers to piece by machine and quilt by hand.

After childhood, Eula Mae stayed away from quilting until 1966. Her reason for beginning again was the expected grandchild-to-be. She just had to make a quilt! Her first attempt was a Sunbonnet Sue and Overall Boy. Eula Mae confesses that this quilt was a nightmare

for her. Her quilting stitches seemed so big. Without her mother to inspect her work and help her improve, Eula Mae got discouraged. She did complete that quilt two weeks after the birth of her grandson, but it took a toll on her quilting interest and enthusiasm. Her husband was instructed to cut her frame up for kindling wood. She said, "I will never again torture myself and my fingers with quilting." Famous last words! For years Eula Mae kept her promise. She did collect quilt patterns, never stitching them, just collecting.

In 1973, Eula Mae and her husband moved to Oregon. She helped form a group of quilters there called "The Capitol Quilters." This group has grown and now has a yearly show with over 200 quilts per show. Eula Mae also belongs to the Northwest Quilters Inc. of Portland, Oregon. Thus, she began quilting again.

Eula Mae's Royal Stars of the States series came about because of her interest in patterns.

She was asked if she had an Oregon Star in her large pattern collection. It bothered her that she didn't. After investigating, she found that many states were without such a pattern. So she began to design. It took her six months to design all fifty of her Royal Stars of the States. She designed and drew the whole quilt diagrams, and her late husband enlarged them and drafted the pieces. She then started to sew them. After completing five of them, she got tired of sewing stars. This was when she got the idea of sharing her designs with others. *Quilt World* magazine began printing the Royal Stars of the States patterns in 1982 and completed the series in 1987.

Eula Mae is delighted to discover so many people making her designs. When Eula Mae learned that Dolores Yoder had completed a quilt from each one of her fifty designs, she was astounded. Eula Mae just had to meet Dolores and see her quilts. This happened in Paducah, Kentucky, in April of 1987. Dolores' exhibition at the Kentucky Oaks Mall drew Eula Mae all the way from Oregon. In Eula Mae's own words, "Words can't really describe my feeling as I walked into that mall and saw all those beautiful quilts made from the designs that I had created. I was speechless and a little numb. It was a feeling I have never felt before."

Now you can share Eula Mae's euphoria with all of the Royal Stars of the States patterns printed together in this book. At the same time you can see Dolores' collection of quilts made with Eula Mae's patterns. We hope to see many more of these quilts hanging in shows around the country and the world in years to come.

Dolores Yoder
Quilter

Dolores Yoder is a Mennonite quilter from Texas. She makes a living selling her quilts! Along the way, and in between her other quilts, Dolores has managed to make all fifty of the Royal Stars of the States quilts. Although Dolores paid other quilters to hand-quilt the tops, she pieced all of these by machine by herself! She fit these in with the three or four quilt tops she finishes in an average week.

Dolores has traveled around the country with her display of the Royal Stars quilts. In Paducah, Kentucky, she realized one of her dreams, to meet the designer of the Royal Stars, Eula Mae Long. What an exciting time for both of them. Dolores' show included some of the other custom quilts she makes to sell, including the all-pieced North Carolina Lily for which she is so well known.

Dolores and her husband, Toby, have six children. Toby travels with her to shows, built her quilt hanging frames and promotes his wife's work wherever they go. Without all of this support, Dolores would not have been able to achieve her goals. Not only does Dolores' immediate family support her, but her mother and her sister sometimes travel with her to help. Such family camaraderie is enviable.

Watching Dolores as she buzzes through her curved seams on her sewing machine makes you wonder if she might have been born knowing how to sew. She makes sewing those curved seams on her quilts look as easy as sewing a straight seam. Her speed and accuracy are overwhelming. One demonstration makes it all look so easy.

For Dolores, the completion of the fifty quilts fulfilled one of her dreams. The display of them at national quilting events fulfilled another. Meeting the designer, Eula Long, topped them all. Dolores had another dream. She had hoped to see all of the Royal Stars of the States patterns together in one book. Now this dream has come true, and she's in it! Dolores says she was led to start this project by seeking the Lord's will, and following His guidance helped her to complete it. We are all happy she did.

Royal Star of
ALABAMA

Where would we be without cotton?! Alabama's proud heritage comes through boldly in this eight-pointed star within a star.

Alabama: The Heart of Dixie
22nd State: December 14, 1819
Capital: Montgomery
Motto: "We Dare to Defend Our Rights"
State Song: "Alabama"
State Flower: Camellia
State Bird: Yellow Hammer
State Tree: Southern Pine

See ALABAMA TEMPLATES on page 62

Royal Star of
ALASKA

Many wondered what Alaska could possibly offer when Russia sold it for $7.2 million. Here is one of its many treasures.

Alaska: The Last Frontier
49th State: January 3, 1959
Capital: Juneau
Motto: "North to the Future"
State Song: "Alaska's Flag"
State Flower: Forget-Me-Not
State Bird: Willow Ptarmigan
State Tree: Sitka Spruce

See ALASKA TEMPLATES on page 65

**See ARIZONA
TEMPLATES
on page 67**

Royal Star of
ARIZONA

Arizona's many natural wonders take your breath
away, just like this magnificent quilt that speaks
of a great state.

Arizona: The Grand Canyon
State
48th State: February 14, 1912
Capital: Phoenix
Motto: "God Enriches"
State Song: "Arizona"
State Flower: Saguaro
State Bird: Cactus Wren
State Tree: Paloverdo

Royal Star of
ARKANSAS

Fine craftsmanship abounds in this state of opportunity, as evidenced in this lovely quilt design.

See
ARKANSAS TEMPLATES
on page 69

Arkansas: The Land of Opportunity
25th State: June 15, 1836
Capital: Little Rock
Motto: "The People Rule"
State Song: "Arkansas"
State Flower: Apple Blossom
State Bird: Mockingbird
State Tree: Pine Tree

Royal Star of
CALIFORNIA

California is a state of diversity. Here, the Golden State shines in all its glory with a bright and colorful quilt.

California: The Golden State
31st State: September 9, 1850
Capital: Sacramento
Motto: "I Have Found It"
State Song: "I Love You California"
State Flower: Golden Poppy
State Bird: California Valley Quail
State Tree: California Redwood

See CALIFORNIA TEMPLATES on page 72

Royal Star of
COLORADO

This majestic quilt represents the many inspiring wonders the Rocky Mountain state offers.

Colorado: The Centennial State
38th State: August 1, 1876
Capital: Denver
Motto: "Nothing Without Providence"
State Song: "Where the Columbines Grow"
State Flower: Columbine
State Bird: Lark Bunting
State Tree: Blue Spruce

See COLORADO TEMPLATES on page 75

See CONNECTICUT TEMPLATES on page 78

Royal Star of
CONNECTICUT

The light of the Constitution State shines through in this brilliantly crafted diamond quilt.

Connecticut: The Constitution State
5th State: January 9, 1788
Capital: Hartford
Motto: "He Who Transplanted Still Sustains"
State Song: None
State Flower: Mountain Laurel
State Bird: Robin
State Tree: White Oak

Royal Star of
DELAWARE

Liberty and independence ring loud and clear in this All-American Quilt of the First State.

Delaware: The First State
1st State: December 7, 1787
Capital: Dover
Motto: "Liberty and Independence"
State Song: "Our Delaware"
State Flower: Peach Blossom
State Bird: Blue Hen Chicken
State Tree: American Holly

See DELAWARE TEMPLATES on page 81

Royal Star of
FLORIDA

Florida: The Sunshine State
27th State: March 3, 1845
Capital: Tallahassee
Motto: "In God We Trust"
State Song: "Swanee River"
State Flower: Orange Blossom
State Bird: Mockingbird
State Tree: Sabal Palm

Feel the life-giving sunshine in all its warmth and brightness in this radiant quilt.

See FLORIDA TEMPLATES on page 83

Royal Star of
GEORGIA

Shining like the stars on a warm
summer evening, this quilt dazzles
the eyes and senses.

See **GEORGIA**
TEMPLATES
on page 87

Georgia: The Empire State of
the South
4th State: January 2, 1788
State Capital: Atlanta
Motto: "Wisdom, Justice and
Moderation"
State Song: "Georgia"
State Flower: Cherokee Rose
State Bird: Brown Thrasher
State Tree: Live Oak

See HAWAII TEMPLATES on page 88

Royal Star of
HAWAII

Favored with nature's blessings, the Aloha State is reflected in this warm and inviting quilt.

Hawaii: The Aloha State
50th State: August 21, 1959
State Capital: Honolulu
Motto: "The Life of the Land Is Perpetuated in Righteousness"
State Song: None
State Flower: Hibiscus
State Bird: Hawaiian Goose
State Tree: Kukui

Royal Star of
IDAHO

Idaho: The Gem State
43rd State: July 3, 1890
Capital: Boise
Motto: "It Is Forever"
State Song: "Here We Have Idaho"
State Flower: Mock Orange
State Bird: Mountain Bluebird
State Tree: Western White Pine

Majestic pines, mountain bluebirds, clean air …
it all comes together in this inspiring quilt.

See IDAHO TEMPLATES on page 90

Royal Star of
ILLINOIS

A great president, a great city, and
now a great quilt from the
Land of Lincoln.

Illinois: The Land of Lincoln
21st State: December 3, 1818
Capital: Springfield
Motto: "State Sovereignty,
    National Union"
State Song: "Illinois"
State Flower: Violet
State Bird: Cardinal
State Tree: White Oak

**See ILLINOIS
TEMPLATES
on page 93**

**See INDIANA
TEMPLATES on
page 95**

95

Royal Star of
INDIANA

Indiana: The Hoosier State	
19th State: December 11, 1816	
Capital: Indianapolis	
Motto: "Crossroads of America"	
State Song: "On the Banks of the Wabash"	
State Flower: Peony	
State Bird: Cardinal	
State Tree: Tulip Tree	

Like stars dancing on the waters of the Wabash, so
sparkles the handiwork of this classic quilt.

Royal Star of
IOWA

Bursting with pride for its fields of plenty, Iowa is represented proudly by this impressive quilt.

See IOWA TEMPLATES on page 98

Iowa: The Hawkeye State
29th State: December 28, 1846
Capital: Des Moines
Motto: "Our Liberties We Prize And Our Rights"
State Song: "The Song of Iowa"
State Flower: Wild Rose
State Bird: Eastern Goldfinch
State Tree: Oak

Royal Star of
KANSAS

Kansas: The Sunflower State
34th State: January 29, 1861
Capital: Topeka
Motto: "To the Stars Through Difficulties"
State Song: "Home on the Range"
State Flower: Sunflower
State Bird: Meadowlark
State Tree: Cottonwood

Like neat rows in vast fields of grain, the Royal Star of Kansas speaks of its abundant and fertile state.

See KANSAS TEMPLATES on page 102

See KENTUCKY TEMPLATES on page 104

Royal Star of
KENTUCKY

Capture the essence of white fences and bluegrass in this quilt, ready-made for an old Kentucky home.

Kentucky: The Bluegrass State
15th State: June 1, 1792
Capital: Frankfort
Motto: "United We Stand, Divided We Fall"
State Song: "My Old Kentucky Home"
State Flower: Goldenrod
State Bird: Cardinal
State Tree: Tulip Poplar

Royal Star of
LOUISIANA

Pretty as a Southern belle, this Royal Star is reminiscent of fragrant magnolias along the bayous.

Louisiana: The Pelican State
18th State: April 30, 1812
Capital: Baton Rouge
Motto: "Justice and
   Confidence"
State Song: "Song of
   Louisiana"
State Flower: Magnolia
State Bird: Brown Pelican
State Tree: Bald Cypress

See LOUISIANA TEMPLATES on page 107

Royal Star of
MAINE

Rugged good looks inspired by an unmatched wilderness mark this beautiful quilt.

Maine: The Pine Tree State
23rd State: March 15, 1820
Capital: Augusta
Motto: "I Direct" or "I Guide"
State Song: "State of Maine"
State Flower: White Pinecone
State Bird: Chickadee
State Tree: White Pine

See MAINE TEMPLATES on page 110

Royal Star of
MARYLAND

Our seventh state is graced with a Royal Star
depicting the strength and resources
of a great people.

Maryland: The Old Line State
7th State: April 28, 1788
Capital: Annapolis
Motto: "Deeds Are Manly,
    Words Are Womanly"
State Song: "Maryland, My
    Maryland"
State Flower: Black-Eyed
    Susan
State Bird: Baltimore Oriole
State Tree: White Oak

**See MARYLAND
TEMPLATES
on page 112**

See MASSACHUSETTS TEMPLATES on page 114

Royal Star of
MASSACHUSETTS

All hail to Massachusetts for its strength
and beauty, and its Royal Star quilt.

Massachusetts: The Bay State
6th State: February 6, 1788
Capital: Boston
Motto: "By The Sword We Seek Peace, But Peace Only Under Liberty"
State Song: "All Hail to Massachusetts"
State Flower: Mayflower
State Bird: Chickadee
State Tree: American Elm

Royal Star of
MICHIGAN

Just look about you and you'll see pleasures far and wide in Michigan. Like this superb quilt.

Michigan: The Wolverine State
26th State: January 26, 1837
Capital: Lansing
Motto: "If You Seek a Pleasant Peninsula, Look About You"
State Song: "Michigan, My Michigan"
State Flower: Apple Blossom
State Bird: Robin
State Tree: White Pine

See **MICHIGAN TEMPLATES** on page 115

Royal Star of
MINNESOTA

Minnesota: The Gopher State
32nd State: May 11, 1858
Capital: St. Paul
Motto: "The Star of the North"
State Song: "Hail! Minnesota"
State Flower: Lady's Slipper
State Bird: Common Loon
State Tree: Norway Pine

Like the bright Star of the North, this Royal quilt shines true and blue for the Gopher State.

See MINNESOTA TEMPLATES on page 117

Royal Star of
MISSISSIPPI

Magnolias and mockingbirds—add this Royal Star
quilt and you've got the magic of Mississippi.

See MISSISSIPPI TEMPLATES on page 118

Mississippi: The Magnolia State
20th State: December 10, 1817
Capital: Jackson
Motto: "By Valor and Arms"
State Song: "Go Mis-sis-sip-pi"
State Flower: Magnolia
State Bird: Mockingbird
State Tree: Magnolia

See MISSOURI TEMPLATES on page 121

Royal Star of
MISSOURI

The Show Me State shows off its Royal Star
in a radiant yet down-to-earth design.

Missouri: The Show Me State
24th State: June 16, 1821
Capital: Jefferson City
Motto: "The Welfare of the
   People Shall Be the
   Supreme Law"
State Song: "Missouri Waltz"
State Flower: Hawthorn
State Bird: Bluebird
State Tree: Flowering Dogwood

Royal Star of
MONTANA

Montana: The Treasure State
41st State: November 8, 1889
Capital: Helena
Motto: "Gold and Silver"
State Song: "Montana"
State Flower: Bitterroot
State Bird: Meadowlark
State Tree: Ponderosa Pine

Gold and silver are often found in the Treasure State. Here's something a bit more rare, a Royal Star.

See MONTANA TEMPLATES on page 123

See NEBRASKA TEMPLATES on page 126

Royal Star of
NEBRASKA

Beautiful Nebraska reveals the splendor of the Cornhusker State in this 12-pointed star.

Nebraska: The Cornhusker State
37th State: March 1, 1867
Capital: Lincoln
Motto: "Equality Before the Law"
State Song: "Beautiful Nebraska"
State Flower: Goldenrod
State Bird: Meadowlark
State Tree: Cottonwood

Royal Star of
NEVADA

Strike it rich in the Silver State
with this eye-catching Royal
Star quilt.

Nevada: The Silver State
37th State: October 31, 1864
Capital: Carson City
Motto: "All for Our Country"
State Song: "Home Means
    Nevada"
State Flower: Sagebrush
State Bird: Bluebird
State Tree: Single-Leaf Pinon

**See NEVADA
TEMPLATES
on page 129**

Royal Star of
NEW HAMPSHIRE

New Hampshire: The Granite State
9th State: June 21, 1788
Capital: Concord
Motto: "Live Free or Die"
State Song: "Old New Hampshire"
State Flower: Purple Lilac
State Bird: Purple Finch
State Tree: White Birch

As granite's strength is found in the depths of the earth—so the charm of this quilt flows from its center.

See NEW HAMPSHIRE TEMPLATES on page 131

Royal Star of
NEW JERSEY

Liberty and prosperity come together in perfect
union in this splendid Royal Star quilt.

See
**NEW JERSEY
TEMPLATES
on page 133**

New Jersey: The Garden State
3rd State: December 18, 1787
Capital: Trenton
Motto: "Liberty and Prosperity"
State Song: None
State Flower: Violet
State Bird: Eastern Goldfinch
State Tree: Red Oak

See NEW MEXICO TEMPLATES on page 136

136

Royal Star of
NEW MEXICO

Feel the rich Indian heritage, the mountains, deserts and bright sunshine in this enchanting quilt.

New Mexico: The Land of Enchantment
47th State: January 12, 1912
Capital: Santa Fe
Motto: "It Grows as It Goes"
State Song: "O, Fair New Mexico"
State Flower: Yucca Flower
State Bird: Roadrunner
State Tree: Nut Pine

Royal Star of
NEW YORK

New York is in the center of things, especially with this magnificent quilt fit for the Empire State.

New York: The Empire State
11th State: July 26, 1788
Capital: Albany
Motto: "Ever Upward"
State Song: None
State Flower: Rose
State Bird: Bluebird
State Tree: Sugar Maple

See NEW YORK TEMPLATES on page 138

See NORTH CAROLINA TEMPLATES on page 140

North Carolina: The Tar Heel State
12th State: November 21, 1789
Capital: Raleigh
Motto: "To Be, Rather Than To Seem"
State Song: "The Old North State"
State Flower: Flowering Dogwood
State Bird: Cardinal
State Tree: Pine

Royal Star of
NORTH CAROLINA

The unchanging charms of a proud state are reflected in this marvelous quilt.

Royal Star of
NORTH DAKOTA

As mellow as a wild prairie rose, this quilt features the hues of the Badlands and its painted rocks.

North Dakota: The Flickertail State
39th State: November 2, 1889
Capital: Bismarck
Motto: "Liberty and Union, Now and Forever, One and Inseparable"
State Song: "North Dakota Hymn"
State Flower: Wild Prairie Rose
State Bird: Meadowlark
State Tree: American Elm

See NORTH DAKOTA TEMPLATES on page 142

See OHIO TEMPLATES on page 144

Royal Star of
OHIO

Beautiful Ohio, the Buckeye State, shows off a stately Royal Star on behalf of its thriving people.

Ohio: The Buckeye State
17th State: October 1, 1803
Capital: Columbus
Motto: "With God All Things Are Possible"
State Song: "Beautiful Ohio"
State Flower: Scarlet Carnation
State Bird: Cardinal
State Tree: Buckeye

Royal Star of
OKLAHOMA

The spirit of pioneering people
and the force of Indian nations
come together in this quilt.

See OKLAHOMA TEMPLATES on page 146

146

Oklahoma: The Sooner State
46th State: November 16, 1907
Capital: Oklahoma City
Motto: "Labor Conquers All Things"
State Song: "Oklahoma"
State Flower: Mistletoe
State Bird: Scissor-Tailed Flycatcher
State Tree: Redbud

See OREGON TEMPLATES on page 148

Royal Star of
OREGON

From the highest peaks to the mighty ocean, Oregon shows its colors in this beautiful quilt.

Oregon: The Beaver State
33rd State: February 14, 1859
Capital: Salem
Motto: "The Union"
State Song: "Oregon, My Oregon"
State Flower: Oregon Grape
State Bird: Western Meadowlark
State Tree: Douglas Fir

Royal Star of
PENNSYLVANIA

Rooted in Old World ways and
traditions, this Royal Star portrays
a proud heritage.

Pennsylvania: The Keystone
State
2nd State: December 12, 1787
Capital: Harrisburg
Motto: "Virtue, Liberty,
Independence"
State Song: None
State Flower: Mountain Laurel
State Bird: Ruffed Grouse
State Tree: Hemlock

**See PENNSYLVANIA
TEMPLATES
on page 152**

Royal Star of
RHODE ISLAND

Rhode Island: Little Rhody
13th State: May 29, 1790
Capital: Providence
Motto: "Hope"
State Song: "Rhode Island"
State Flower: Violet
State Bird: Rhode Island Red
State Tree: Red Maple

Bursting with energy and vitality, Little Rhody shows its big heart in this Royal Star quilt.

See RHODE ISLAND TEMPLATES on page 154

Royal Star of
SOUTH CAROLINA

Radiant with color and excitement, this Royal Star showers its brilliance upon the Palmetto State.

South Carolina: The Palmetto State
8th State: May 23, 1788
Capital: Columbia
Motto: "Prepared in Mind and Resources, While I Breath, I Hope"
State Song: "Carolina"
State Flower: Carolina Jessamine
State Bird: Carolina Wren
State Tree: Palmetto

See SOUTH CAROLINA TEMPLATES on page 156

Royal Star of
SOUTH DAKOTA

The Sunshine State rises to new heights in quilting with this vivid Royal Star quilt.

South Dakota: Sunshine State
40th State: November 2, 1889
Capital: Pierre
Motto: "Under God the People Rule"
State Song: "Hail South Dakota"
State Flower: American Pasque
State Bird: Ring-Necked Pheasant
State Tree: Black Hills Spruce

See SOUTH DAKOTA TEMPLATES on page 158

Royal Star of
TENNESSEE

The hidden fire of black coal comes through in this impressive quilt, perfect for any home.

See TENNESSEE TEMPLATES on page 160

160

Tennessee: The Volunteer State
16th State: June 1, 1796
Capital: Nashville
Motto: "Agriculture and Commerce"
State Song: "The Tennessee Waltz"
State Flower: Iris
State Bird: Mockingbird
State Tree: Tulip Poplar

**See TEXAS
TEMPLATES
on page 164**

Texas: The Lone Star State
28th State: December 29, 1845
Capital: Austin
Motto: "Friendship"
State Song: "Texas, Our Texas"
State Flower: Bluebonnet
State Bird: Mockingbird
State Tree: Pecan

Royal Star of

TEXAS

The wide open spaces of the Lone Star State
echo in this magnificent Royal Star quilt.

Royal Star of
UTAH

Broad valleys, industrious people and a beautiful
Royal Star quilt grace the Beehive State.

**See UTAH
TEMPLATES
on page 166**

Utah: The Beehive State
45th State: January 4, 1896
Capital: Salt Lake City
Motto: "Industry"
State Song: "Utah, We Love
Thee"
State Flower: Sego Lily
State Bird: Sea Gull
State Tree: Blue Spruce

See VERMONT TEMPLATES on page 168

Royal Star of
VERMONT

Color beyond compare, like that of the Green Mountain State, comes through in this attractive quilt.

Vermont: Green Mountain State
14th State: March 4, 1791
Capital: Montpelier
Motto: "Freedom and Unity"
State Song: "Hail Vermont"
State Flower: Red Clover
State Bird: Hermit Thrush
State Tree: Sugar Maple

Royal Star of
VIRGINIA

Unforgettable landscapes and this beautiful
Royal Star quilt carry you back to Old Virginia.

Virginia: The Mother of Presidents
10th State: June 25, 1788
Capital: Richmond
Motto: "Thus Always to Tyrants"
State Song: "Carry Me Back to Old Virginny"
State Flower: Flowering Dogwood
State Bird: Cardinal
State Tree: None

See VIRGINIA TEMPLATES on page 170

**See
WASHINGTON
TEMPLATES
on page 173**

Royal Star of
WASHINGTON

Wooden slopes make this the Evergreen State.
A radiant star makes this an ever-popular quilt.

Washington: The Evergreen State
42nd State: November 11, 1889
Capital: Olympia
Motto: "Bye and Bye"
State Song: "Washington, My
Home"
State Flower: Rhododendron
State Bird: Goldfinch
State Tree: Western Hemlock

Royal Star of
WEST VIRGINIA

Mountaineers are always free—likewise the spirit of this dynamic Royal Star of the Mountain State.

West Virginia: The Mountain State
35th State: June 20, 1863
Capital: Charleston
Motto: "Mountaineers Are Always Free"
State Song: "The West Virginia Hills"
State Flower: Rhododendron
State Bird: Cardinal
State Tree: Sugar Maple

See **WEST VIRGINIA TEMPLATES on page 177**

Royal Star of
WISCONSIN

Stars dance on a thousand clear lakes in this brilliant quilt of the Badger State.

Wisconsin: The Badger State
30th State: May 29, 1848
Capital: Madison
Motto: "Forward"
State Song: "On, Wisconsin"
State Flower: Wood Violet
State Bird: Robin
State Tree: Sugar Maple

See WISCONSIN TEMPLATES on page 178

See WYOMING TEMPLATES on page 180

180

Wyoming: The Equality State
44th State: July 10, 1890
Capital: Cheyenne
Motto: "Equal Rights"
State Song: "Wyoming"
State Flower: Indian Paintbrush
State Bird: Meadowlark
State Tree: Cottonwood

Royal Star of
WYOMING

Majestic peaks of grand mountain ranges echo their glory in this Royal Star quilt.

Royal Stars
General Instructions

This book is not intended to be a book to teach you to make quilts or to encourage you to try new methods. It contains a collection of star patterns and a few coordinating border designs. We make the assumption that you have mastered the basics of quiltmaking and are looking for a challenge. Our basic instructions do not include detailed how-to's or hints for fast-piecing techniques, although they may be used with many of the patterns.

Making star patterns can be very easy or very hard, depending on how accurately you cut your pieces and sew your seams. These Royal Star patterns, much like the more familiar *Lone Star* design, all radiate from the center to the outside. Some of the designs have only eight points, while others have 12 or more. Some of them have only two or three templates, while others have 10 or more.

Fabric yardage is given for each quilt using the colors shown in the photos of Dolores Yoder's finished quilts. These colors are not intended to be the official colors for any state. Substitute your favorite colors in place of hers and adjust the fabric list accordingly.

The patterns we give here do not include a 1/4" seam allowance. To begin making a quilt, prepare templates using the patterns given. To do this, transfer the pattern shape to a sturdy template material, either cardboard or purchased template plastic.

If you will be machine-piecing, add a 1/4" seam allowance to each template before cutting it out. If hand-piecing, cut templates as given and add the seam allowance when cutting fabric pieces.

To cut fabric, lay the template down on the wrong side of one single layer of fabric. Trace around it with a sharp pencil. If hand-piecing, this line will be the sewing line. Leave room enough between pieces for a seam allowance around each piece when cutting.

Grainline suggestions are given on each pattern piece. Place the piece with the arrow parallel to the lengthwise thread of the fabric for best results. Mark and cut the number of pieces needed, reversing piece when directed.

To test the pattern and your stitching, you may want to cut enough pieces to complete only one unit or section of the chosen design. If there is a problem, you will find it in this test section and can solve it before you get into piecing the whole quilt.

Sew pieces together referring to the large piecing diagram and the instructions given with each quilt. Most the the designs are pieced in sections or star points. The points are joined and the fill-in pieces are added to square up the design.

To sew, pin pieces together at matching point; sew from pin to pin. Hand-piecing is recommended; and stitching to the end of the seam, not to the end of the piece, is preferred. This method leaves the points open for setting in pieces.

Many of the star designs are pieced with the B diamond, which is easily pieced in rows. The large drawings will show you how many B pieces are needed in each row to complete the design.

The star designs resulting in eight points all have fill-in triangles and squares to square up the top. Those squares and triangles are given as measurements, not as templates, because of their size. The measurements given are the finished size—add a 1/4" seam allowance to *all* measurements before cutting.

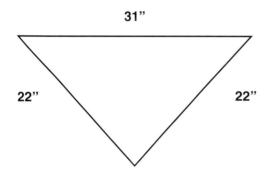

Several pieces are common to many patterns. These templates are given on the following pages with alphabetical labels instead of the numbered labels on the pieces given with each individual state's design. Therefore, when constructing the Royal Star of New Jersey, for example, you are told on page 133 to cut a total of 224 B pieces in various colors and are referred to page 61. B is a common template and is not with the other New Jersey pattern pieces.

Many of the Royal Star designs begin in the center with a round template and work out to a round design that must be squared off to complete the top. These designs use piece A for that outside piece.

Piece A may be cut larger than the instructions direct if you want a larger quilt. If the piece is cut as directed, you will notice that the point almost disappears as it meets the star point at the center edge. If you cut the piece larger, you will have a seam where the A pieces meet at this point. It is better to have the piece larger as you can always cut away any excess to square up the quilt top when finished.

It is easier to appliqué round center pieces on than it is to piece them in. If you prefer piecing, be careful to make the center perfectly round.

Due to space limitations, we have selected only a few border patterns to share with you. A couple are found with the star patterns, but the remainder are found starting on page 183. More information about adding borders is given in that section.

When the quilt top is finished and borders have been added, you will need to choose a quilting design. Because many of the stars have large open spaces, there is room to show off your quilting design and best stitches. There are many quilt-design books available to supply you with designs to enhance your quilt. After the design is chosen, mark it on the top using a light chalk pencil or a water-soluble marking pen.

Prepare a backing for your quilt at least 4" larger all around. Sandwich the chosen batting between the prepared backing and the completed top. Baste the layers together to hold flat for quilting. Quilt as desired on marked lines and in seams.

When the quilting is complete, trim edges even and bind with either self-made or purchased binding to finish the edges.

Your quilt is now ready to adorn a bed or wall in a special place in your home.

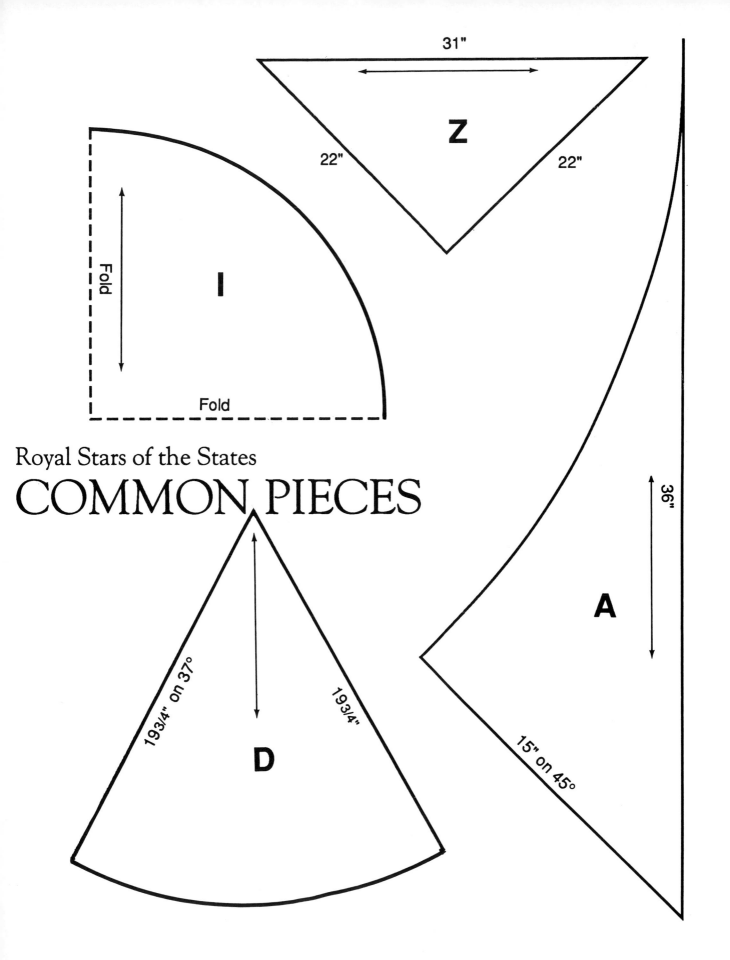

Royal Stars of the States
COMMON PIECES

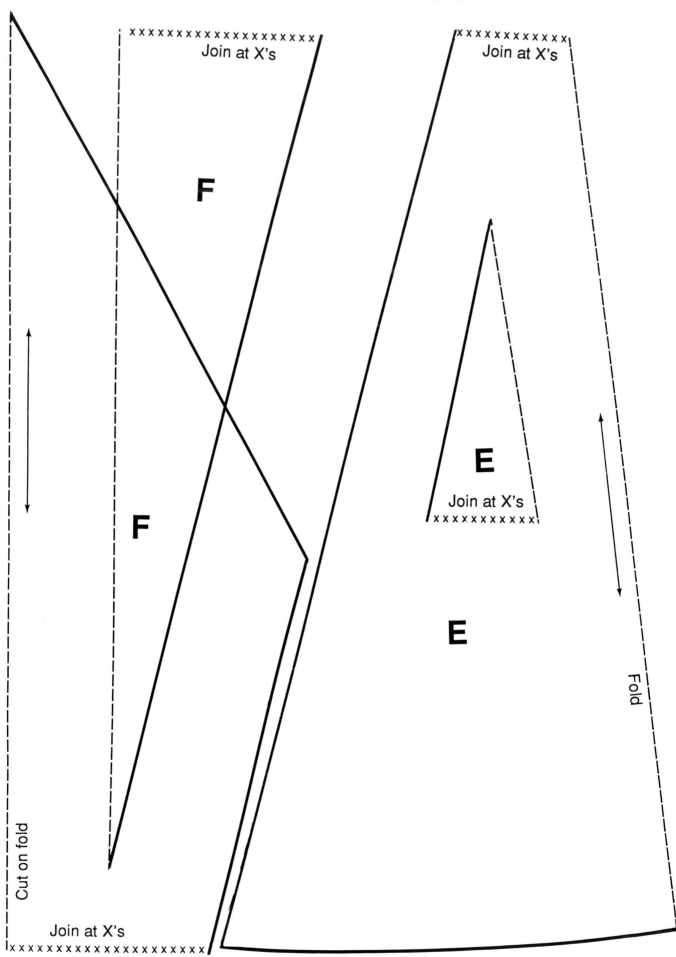

Join at X's

Join at X's

F

F

Join at X's

E

E

Join at X's

Cut on fold

Fold

COMMON PIECES

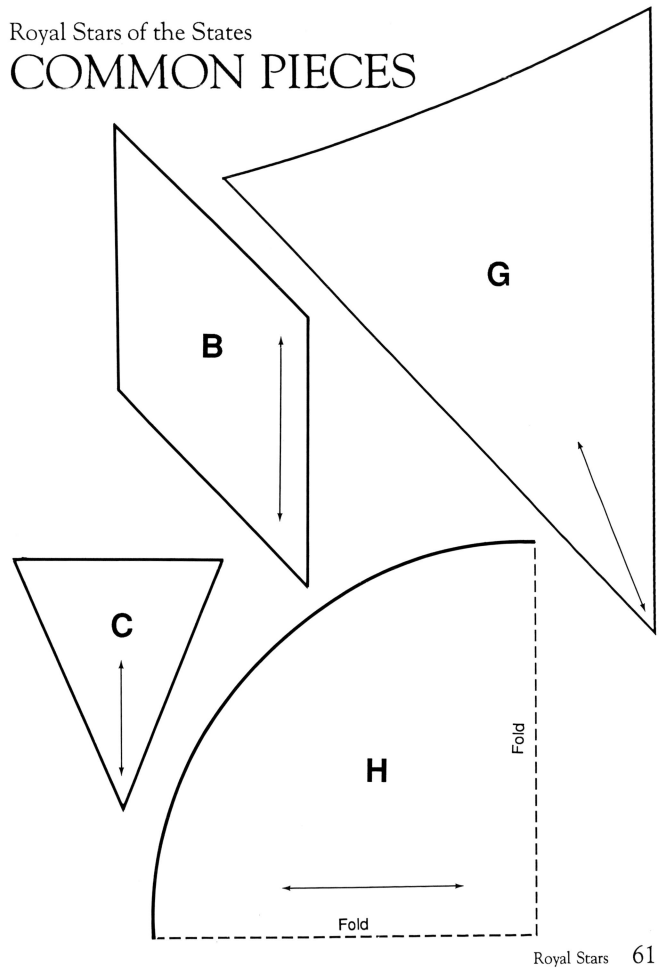

B

G

C

H

Fold

Fold

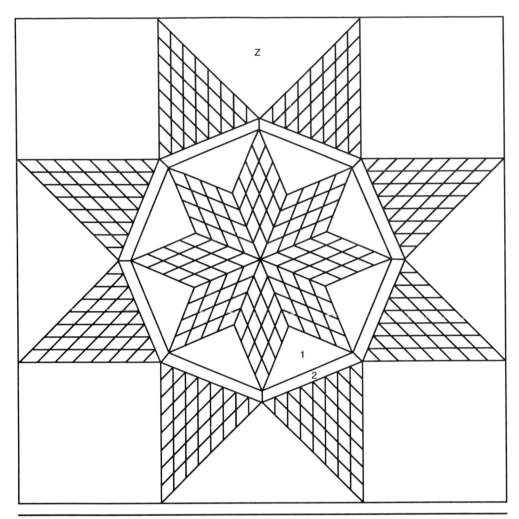

75" x 75"

MATERIALS

- 1/2 yard brown solid
- 3/4 yard white print
- 3/4 yard brown print
- 1/2 yard dark blue print
- 3/4 yard dark blue solid
- 3/4 yard light blue solid
- 3 yards white solid

PIECING INSTRUCTIONS

1. Add a 1/4" seam allowance to all pieces when cutting.

2. Cut four corner squares, each 22" (add seams), from solid white.

3. Cut four Z fill-in triangles (add seams).

4. Cut all other pieces following instructions on piece.

5. To piece the star portion, sew two small diamonds together as shown in Figure #1.

Royal Star of ALABAMA

See ALABAMA photo on page 7

6. Piece four diamonds together to form a row as shown in Figure #2. The rows are joined together to form a larger diamond (Figure #3).

7. Sew two of these diamonds together; make three more the same way.

8. Sew two of these sections together to form half of the star. Sew next two exactly the same to form the center star.

9. When star is pieced together, set in #1 triangles; attach #2 bars to form an octagon.

10. Sew large half diamonds or star points in same manner, except start row with small C triangle instead of the B diamond.

11. To finish the quilt center, sew in large squares and Z fill-in triangles; add borders using a pattern from the border section as desired.

More Alabama
Templates Next Page

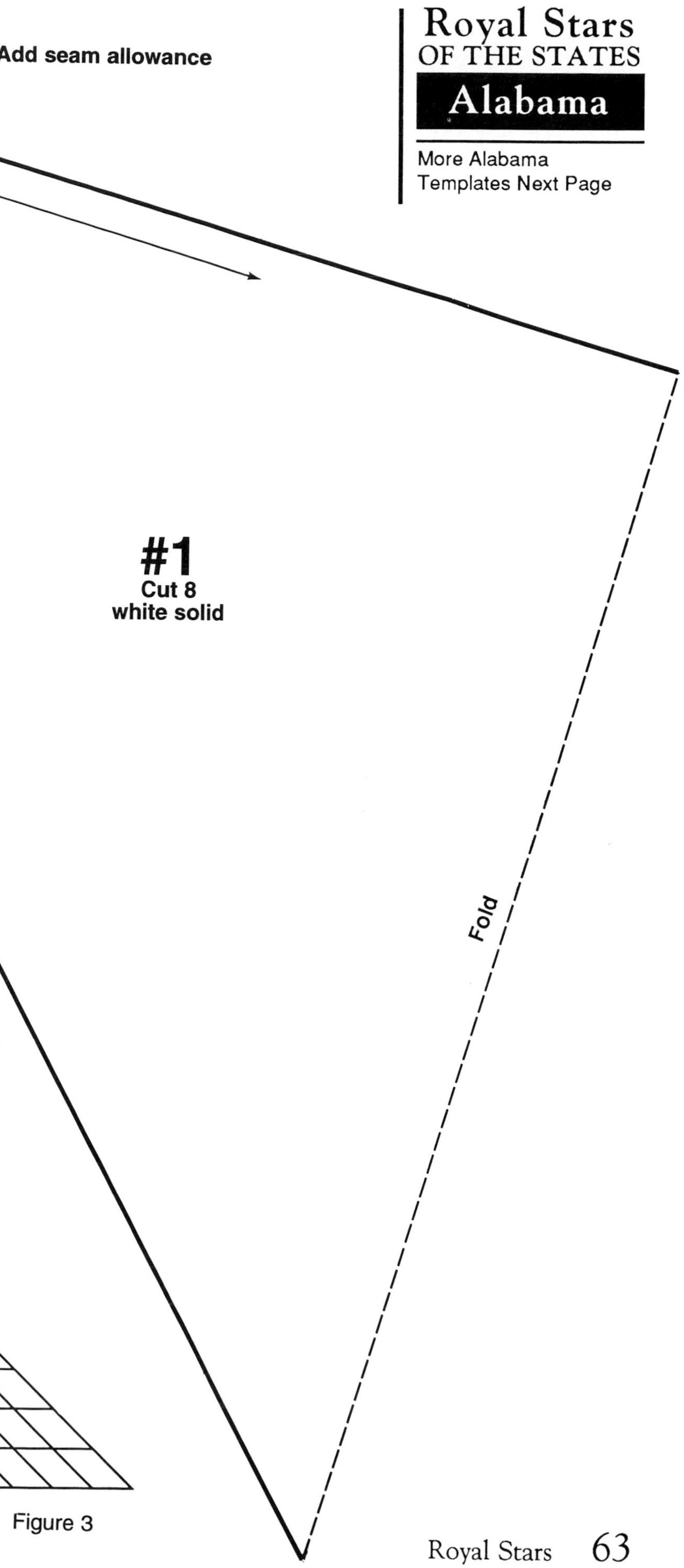
Add seam allowance
#1
Cut 8
white solid
Fold

ADD
COMMON PIECES **B, C & Z**
(REFER TO PAGES 59-61)

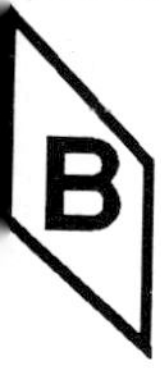
B

Cut 72 white print, 80 light blue solid, 64 brown print, 48 dark blue print, 80 dark blue solid & 8 brown solid

Cut 64 brown solid

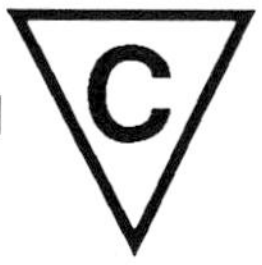
C

Cut 4

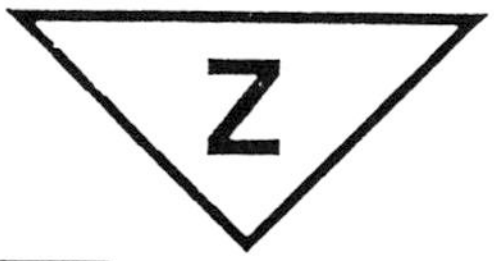
Z

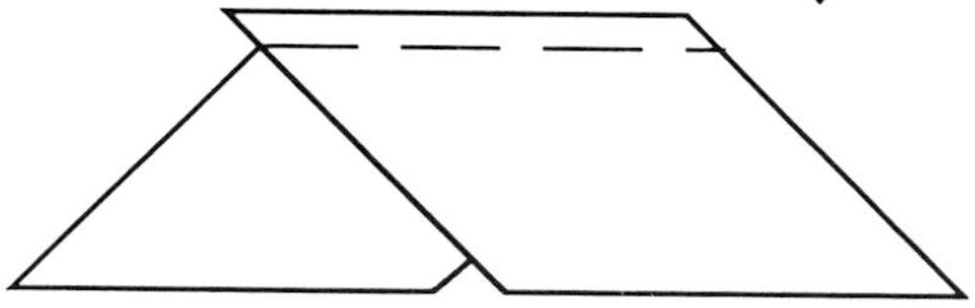

Figure 1

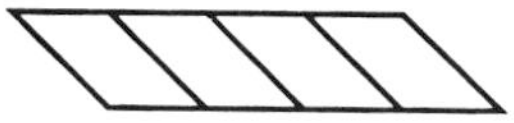

Figure 2

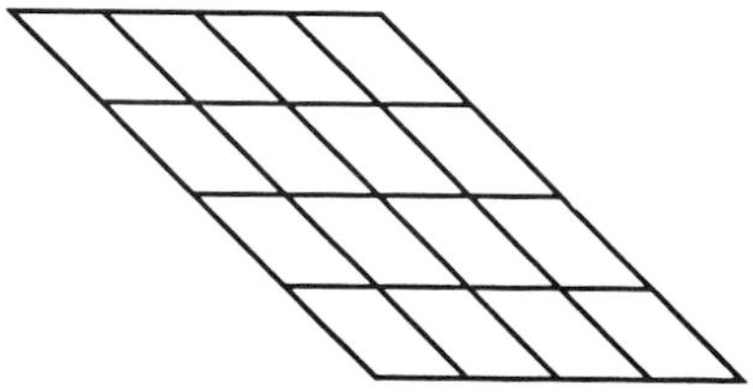

Figure 3

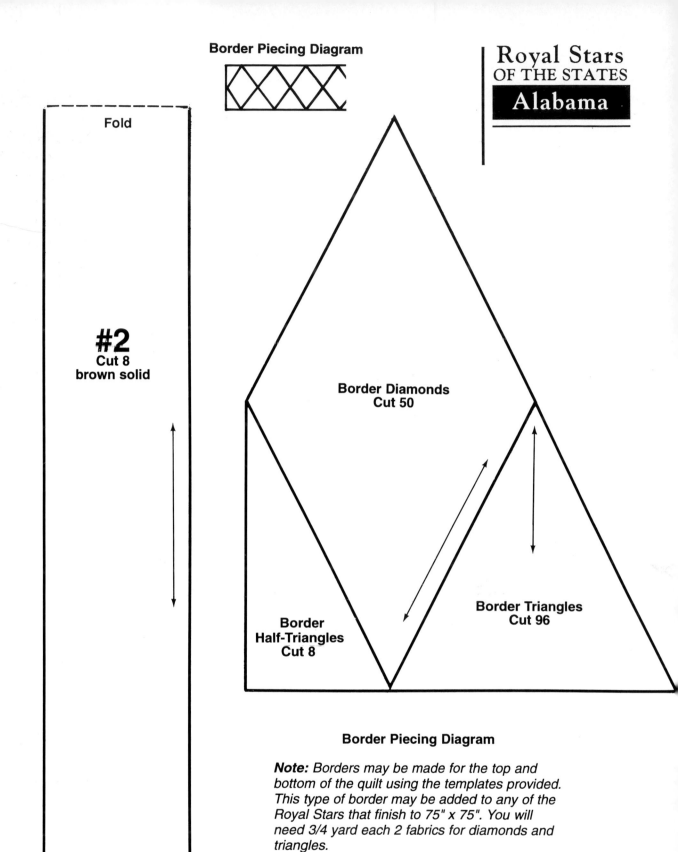

Border Piecing Diagram

#2
Cut 8
brown solid

Fold

Border Diamonds
Cut 50

Border
Half-Triangles
Cut 8

Border Triangles
Cut 96

Border Piecing Diagram

Note: *Borders may be made for the top and bottom of the quilt using the templates provided. This type of border may be added to any of the Royal Stars that finish to 75" x 75". You will need 3/4 yard each 2 fabrics for diamonds and triangles.*

Add seam allowance

75" x 75"

MATERIALS
- 1/2 yard pink
- 2 1/4 yards dark print
- 3/8 yard rust solid
- 1/4 yard dark blue print
- 1/4 yard blue solid
- 1/4 yard light blue print
- 1/4 yard medium blue print
- 1/2 yard purple solid
- 3 yards white solid

PIECING INSTRUCTIONS

1. Cut four corner squares, each 22" (add seams), from solid white.

2. Cut four Z fill-in triangles (add seams).

3. Cut all other pieces from templates (add seam allowance).

4. Piece four small center B triangles together as shown in Figure #1.

Royal Star of ALASKA

5. Piece center section with #1, #2 and B as shown in Figure #2.

6. Sew #1 bars to center diamond as shown in Figure #3.

7. Sew second set of #1 bars to center sections as shown in Figure #4.

8. Sew the sections made in step 6 to those made in step 7.

9. Sew the long #3 and #4 bars and B diamonds onto outer edge to form a large diamond or star point.

10. Piece all star points in the same manner; sew together to form an eight-pointed star.

11. Set in large squares and triangles; finish with borders if desired.

See ALASKA photo on page 8

ADD
COMMON PIECES B & Z
(REFER TO PAGES 59–61)

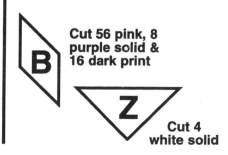

Cut 56 pink, 8 purple solid & 16 dark print

Cut 4 white solid

Add seam allowance

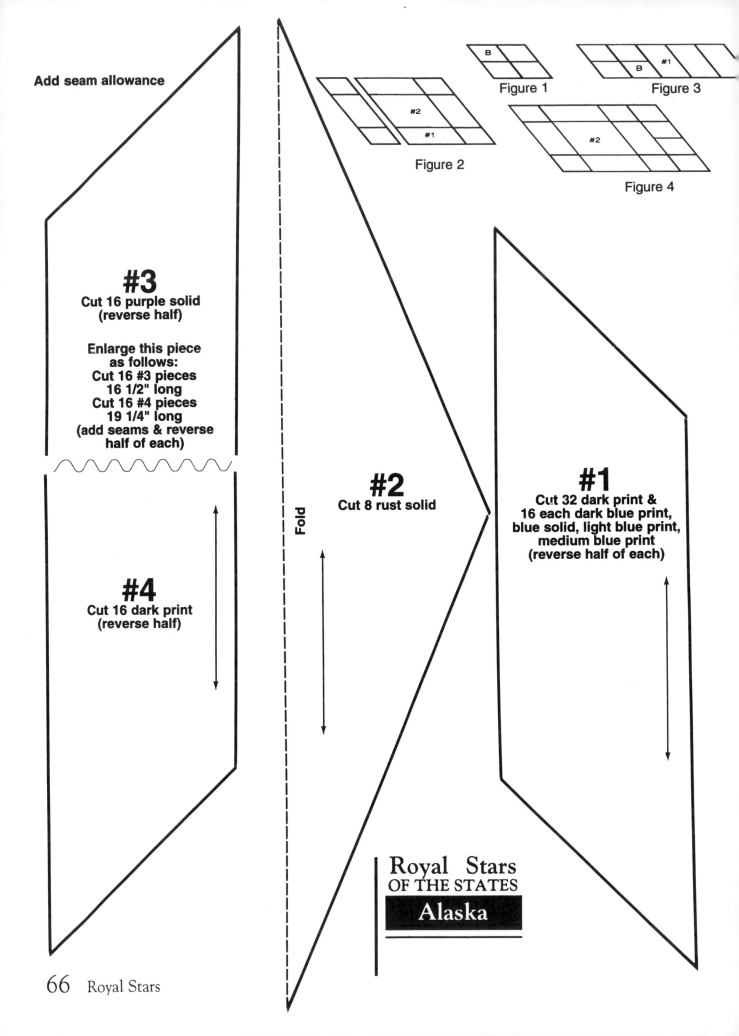

Figure 1

Figure 3

Figure 2

Figure 4

#3
Cut 16 purple solid
(reverse half)

Enlarge this piece
as follows:
Cut 16 #3 pieces
16 1/2" long
Cut 16 #4 pieces
19 1/4" long
(add seams & reverse
half of each)

#4
Cut 16 dark print
(reverse half)

Fold

#2
Cut 8 rust solid

#1
Cut 32 dark print &
16 each dark blue print,
blue solid, light blue print,
medium blue print
(reverse half of each)

Royal Stars
OF THE STATES
Alaska

66" x 66"

MATERIALS
- 1/8 yard pink #1
- 1/2 yard pink #2
- 1 yard pink #3
- 3/4 yard pink #4
- 1/2 yard pink #5
- 1 yard brown
- 3 yards white solid

PIECING INSTRUCTIONS

1. Cut four 21 1/2" squares from white solid for corners (add seams).

2. Cut four fill-in side triangles, enlarging shape as shown on piece #2.

3. Piece diamonds together to make a complete row as shown in Figure #2.

4. Piece five rows in the same manner; sew these rows together to form one large diamond.

5. Sew piece #1 onto this diamond; sew B to the end of the reverse #1 piece and sew to opposite side as shown in Figure #1. Repeat for eight large diamonds.

6. Sew the large diamonds together to form a star.

7. Set in white corner squares and #2 triangles.

8. Sew seven B pieces together and appliqué to the top of each diamond or star point to form the small stars at the tips of each diamond.

9. Add borders if desired.

Royal Star of ARIZONA

See ARIZONA photo on page 9

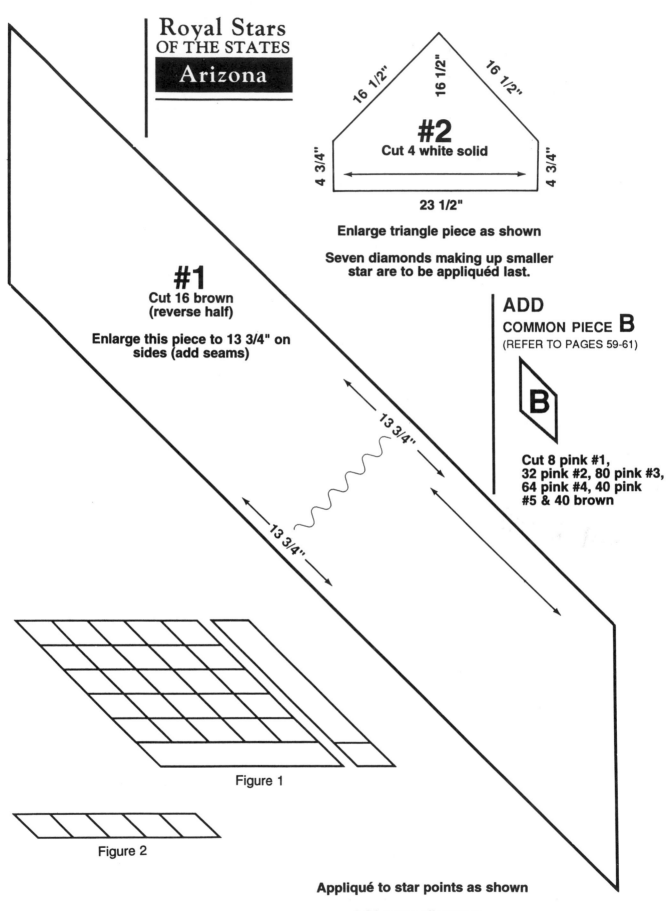

Royal Stars
OF THE STATES
Arizona

16 1/2"   16 1/2"   16 1/2"

#2
Cut 4 white solid

4 3/4"   4 3/4"

23 1/2"

Enlarge triangle piece as shown

Seven diamonds making up smaller star are to be appliquéd last.

#1
Cut 16 brown
(reverse half)

Enlarge this piece to 13 3/4" on sides (add seams)

ADD
COMMON PIECE **B**
(REFER TO PAGES 59-61)

B

Cut 8 pink #1, 32 pink #2, 80 pink #3, 64 pink #4, 40 pink #5 & 40 brown

13 3/4"

13 3/4"

Figure 1

Figure 2

Appliqué to star points as shown

Add seam allowance

72" x 81"

MATERIALS
- 1 yard pink print
- 3 1/2 yards pink solid
- 1/4 yard white print
- 1 1/2 yards dark print
- 2 1/2 yards white solid

PIECING INSTRUCTIONS

Note: *Pieces #2, #3, #4, #5 and #6 must be enlarged before cutting. Refer to measurements given with patterns for enlarging. Add a seam allowance when cutting pieces.*

1. Sew #9 to #8 to #7; repeat for reverse pieces. Sew together to form a star point. Repeat for 12 points. Sew these units together to form a medallion center.

2. Appliqué I on the center of the star.

3. Set #5 and #6 pieces into the star points.

Royal Star of ARKANSAS

See ARKANSAS photo on page 10

4. Set in #4 pieces.

5. Sew a #1 piece to each #3 piece; sew two of these #3 pieces together. Add #2 to this section; repeat for six units. Sew a unit to each side of the hexagon star center.

6. Sew the last three #1 pieces together; set in place to form stars.

7. Sew #11 pieces on the large hexagon to complete the top.

8. Add borders if desired.

ADD
COMMON PIECE I
(REFER TO PAGES 59–61)

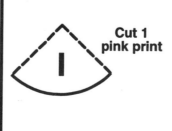

Cut 1
pink print

I

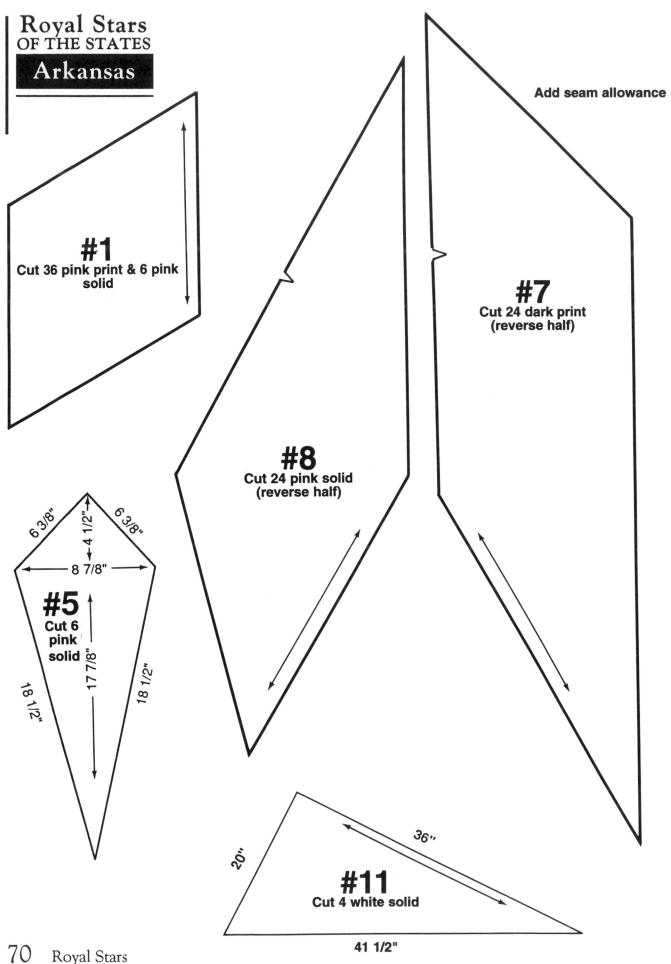

Royal Stars
OF THE STATES
Arkansas

Add seam allowance

#1
Cut 36 pink print & 6 pink solid

#7
Cut 24 dark print (reverse half)

#8
Cut 24 pink solid (reverse half)

6 3/8"   4 1/2"   6 3/8"
8 7/8"

#5
Cut 6 pink solid

17 7/8"

18 1/2"   18 1/2"

#11
Cut 4 white solid

20"   36"

41 1/2"

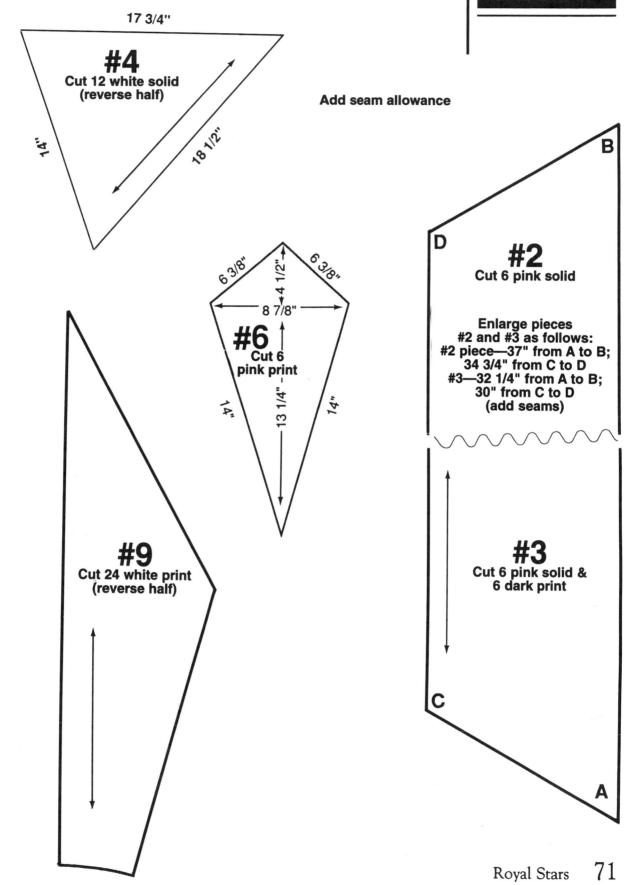

17 3/4"

#4
Cut 12 white solid
(reverse half)

14"

18 1/2"

Add seam allowance

B

D

#2
Cut 6 pink solid

Enlarge pieces
#2 and #3 as follows:
#2 piece—37" from A to B;
34 3/4" from C to D
#3—32 1/4" from A to B;
30" from C to D
(add seams)

6 3/8"   4 1/2"   6 3/8"

8 7/8"

#6
Cut 6
pink print

14"   13 1/4"   14"

#9
Cut 24 white print
(reverse half)

#3
Cut 6 pink solid &
6 dark print

C

A

72" x 72"

MATERIALS
- 2/3 yard blue print
- 1 1/4 yards yellow print
- 1/3 yard red print
- 2/3 yard green print
- 4 1/4 yards white solid

PIECING INSTRUCTIONS

1. Sew #8, #7, #6, #5, #4 and #3 together in the sequence given. Repeat for reverse pieces.

2. Sew two groups of these (one in reverse) together to form one point of the 12-pointed star. Repeat for 12 points.

3. When all star points are complete, sew together to form a star.

4. Appliqué #9 in the center.

5. Set D pieces in between star points to make a circle.

6. Add corner A pieces to complete.

7. Add borders if desired.

Royal Star of CALIFORNIA

See CALIFORNIA photo on page 11

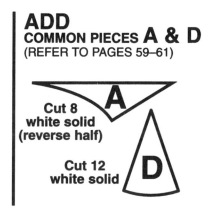

ADD
COMMON PIECES A & D
(REFER TO PAGES 59–61)

A
Cut 8 white solid (reverse half)

D
Cut 12 white solid

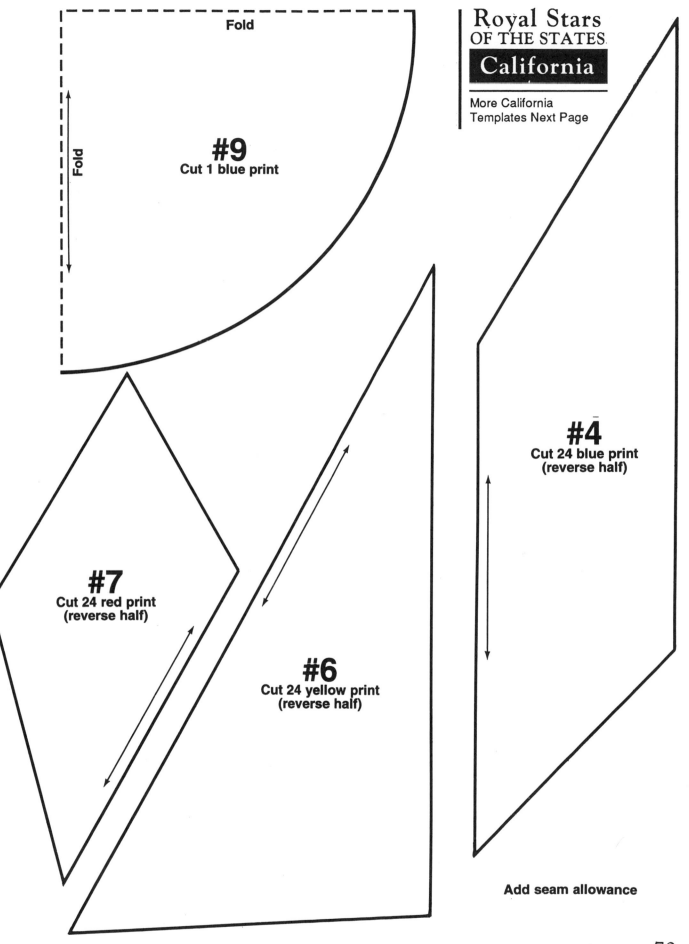

Fold

Fold

#9
Cut 1 blue print

Royal Stars
OF THE STATES.
California

More California
Templates Next Page

#4
**Cut 24 blue print
(reverse half)**

#7
**Cut 24 red print
(reverse half)**

#6
**Cut 24 yellow print
(reverse half)**

Add seam allowance

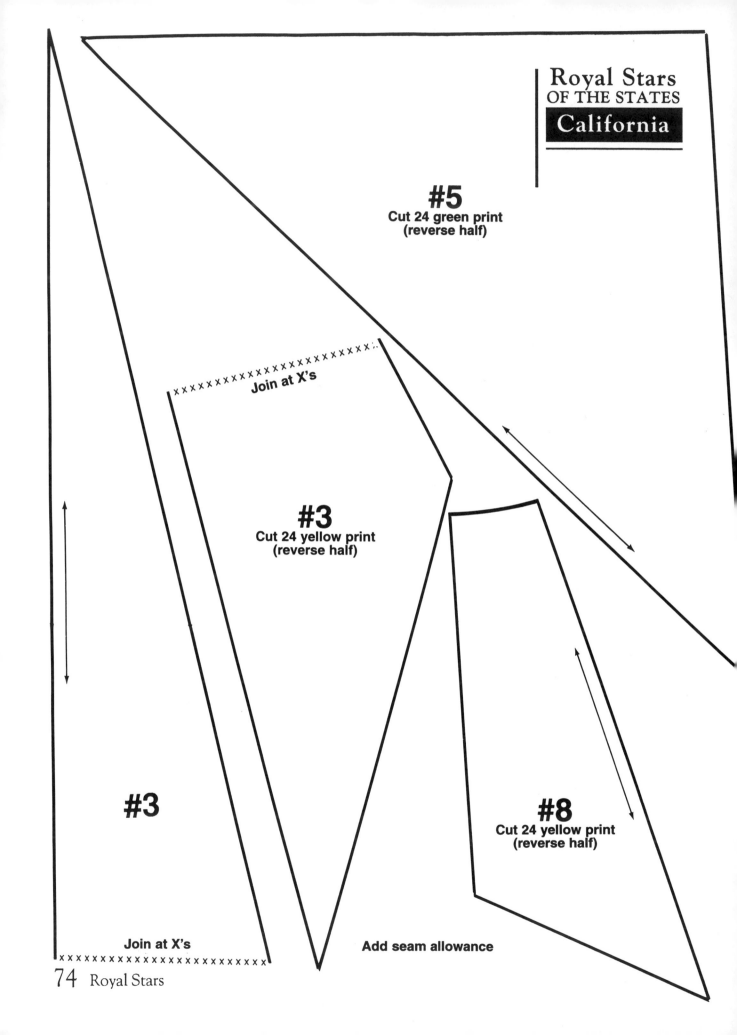

Royal Stars
OF THE STATES
California

#5
Cut 24 green print
(reverse half)

Join at X's

#3
Cut 24 yellow print
(reverse half)

#3

#8
Cut 24 yellow print
(reverse half)

Join at X's

Add seam allowance

72" x 72"

MATERIALS
- 1/3 yard rust
- 1/4 yard blue
- 1/4 yard lavender
- 1/3 yard light blue
- 1/2 yard beige
- 1 yard cream
- 1 3/4 yards gray
- 3 1/2 yards white solid

PIECING INSTRUCTIONS

1. Sew pieces #9, #8, #7, #6, #5 and #4 together in the sequence given. Repeat for reverse pieces.

2. Sew the two sections together to form a star point. Repeat for 12 star points.

3. Sew the 12 points together to form the center star.

4. Appliqué H in place.

5. Sew E to each side of F. Repeat for all E-F pieces; set in between star points.

6. Sew on all A's to finish top.

7. Add borders if desired.

Royal Star of COLORADO

See COLORADO photo on page 12

ADD
COMMON PIECES A, E, F & H
(REFER TO PAGES 59–61)

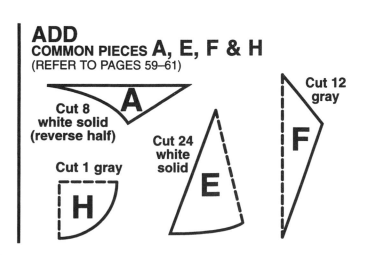

Cut 8 white solid (reverse half)

A

Cut 1 gray

H

Cut 24 white solid

E

Cut 12 gray

F

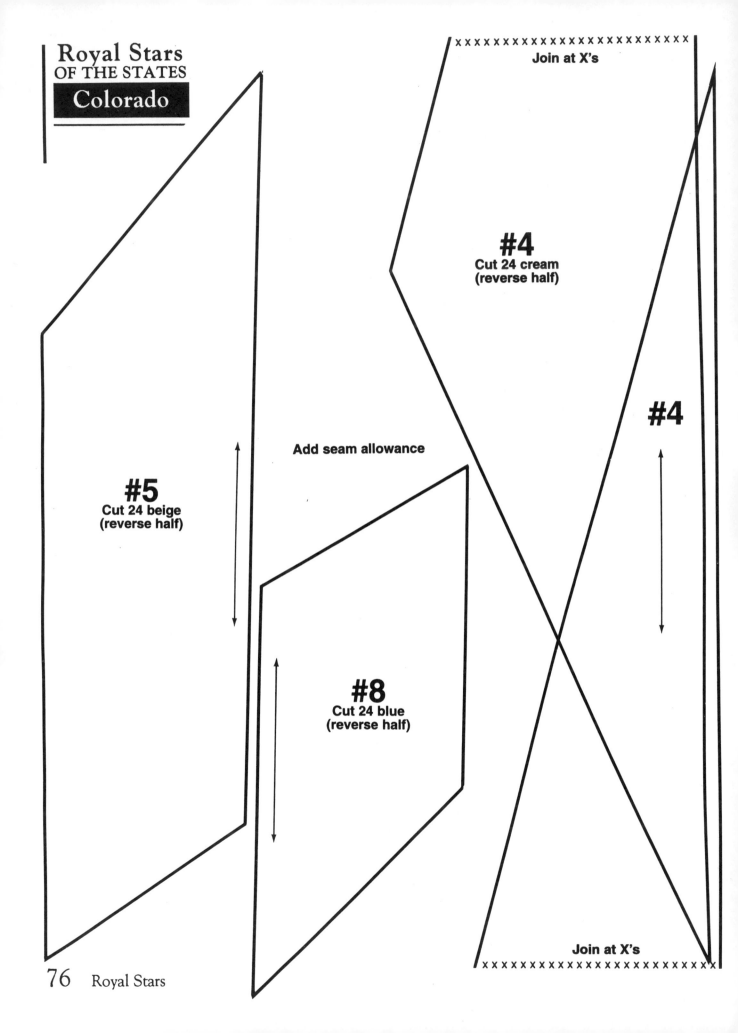

Royal Stars
OF THE STATES
Colorado

Join at X's

#4
Cut 24 cream
(reverse half)

#4

Add seam allowance

#5
Cut 24 beige
(reverse half)

#8
Cut 24 blue
(reverse half)

Join at X's

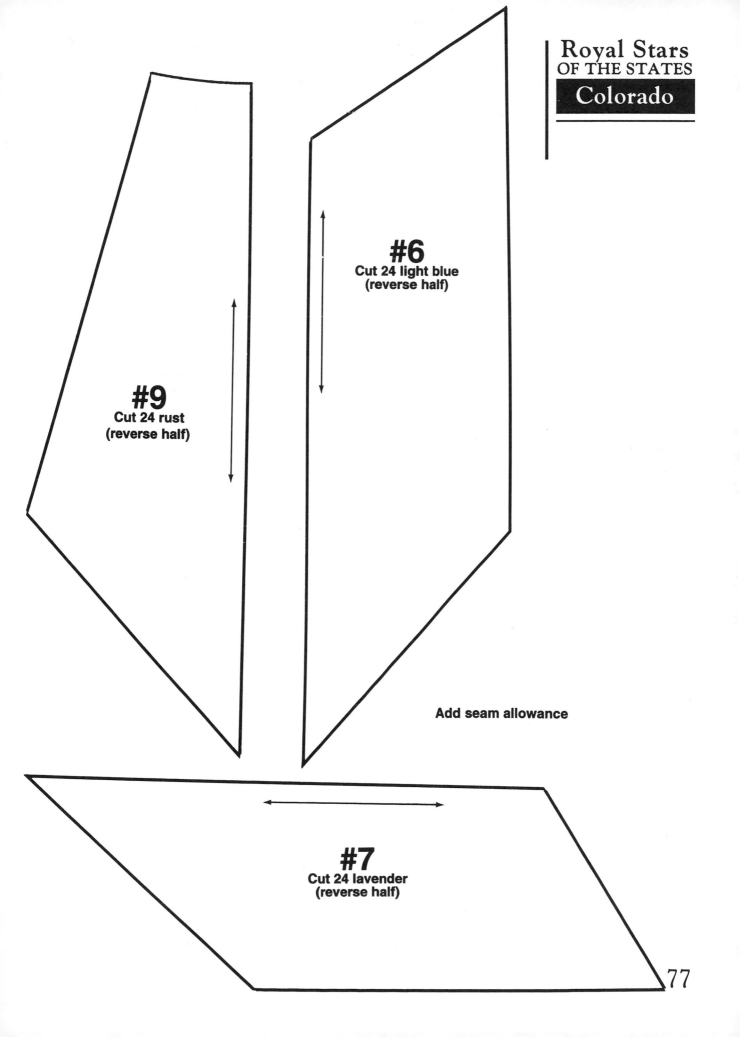

#6
Cut 24 light blue
(reverse half)

#9
Cut 24 rust
(reverse half)

Add seam allowance

#7
Cut 24 lavender
(reverse half)

72" x 72"

MATERIALS
- 1 1/4 yards rust solid
- 1 3/4 yards navy solid
- 1 1/4 yards white print
- 1/3 yard light blue solid
- 1 yard medium blue print
- 3 yards white solid

PIECING INSTRUCTIONS

Note: *When making this type of quilt, start in the center and add each row or number as you work outward. This star is one of the harder ones to construct, but is worth the effort since it results in such a beautiful quilt.*

1. Sew all #11 pieces together to form the center; appliqué H on top. **Note:** *Photo shows I appliquéd on H.*

2. Set in pieces #10 and #9.

3. Sew pieces #6, #7 and #8 together. Repeat for reverse pieces and join. Set units in between the #9 pieces.

4. Sew #3, #4 and #5 together. Repeat for reverse pieces and join. Set units in between the #6 pieces.

5. Set E between #3 points to finish circle.

6. Sew A pieces to outside edges to make square.

7. Add borders if desired.

Royal Star of CONNECTICUT

See CONNECTICUT photo on page 13

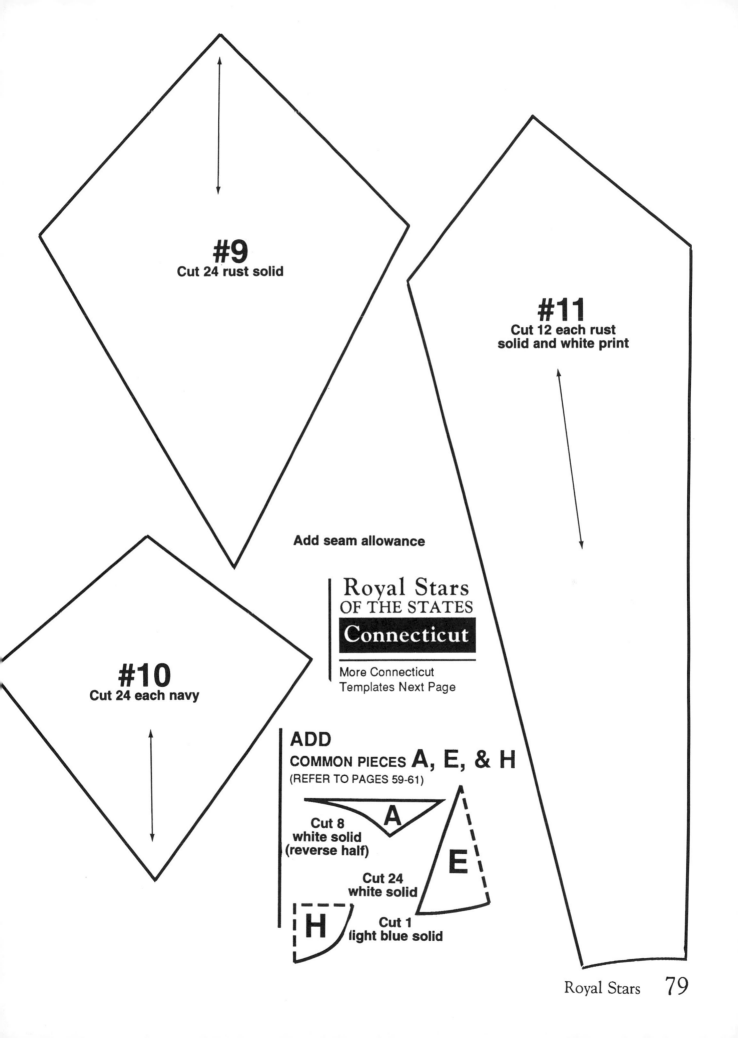

#9
Cut 24 rust solid

#11
Cut 12 each rust
solid and white print

Add seam allowance

Royal Stars
OF THE STATES
Connecticut

More Connecticut
Templates Next Page

#10
Cut 24 each navy

ADD
COMMON PIECES A, E, & H
(REFER TO PAGES 59-61)

A
Cut 8
white solid
(reverse half)

E
Cut 24
white solid

H
Cut 1
light blue solid

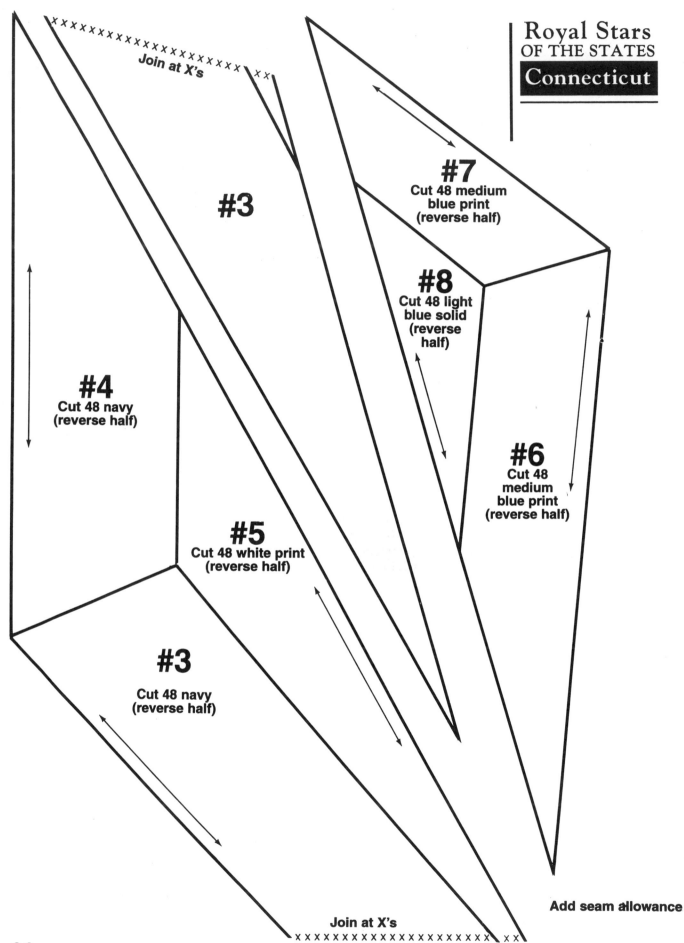

Join at X's

#3

#7
Cut 48 medium
blue print
(reverse half)

#8
Cut 48 light
blue solid
(reverse
half)

#4
Cut 48 navy
(reverse half)

#5
Cut 48 white print
(reverse half)

#6
Cut 48
medium
blue print
(reverse half)

#3
Cut 48 navy
(reverse half)

Add seam allowance

Join at X's

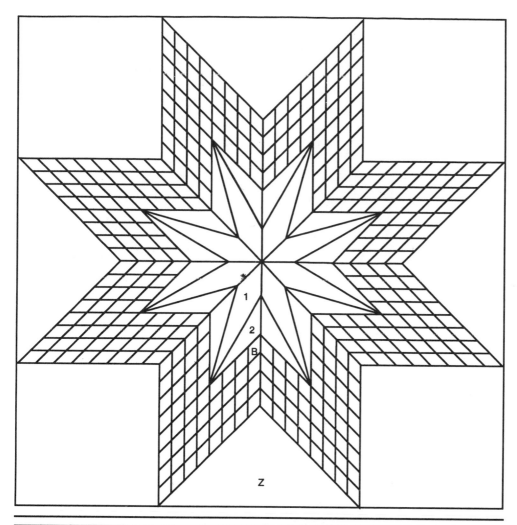

75" x 75"

MATERIALS
- 2/3 yard light lavender
- 3/4 yard lighter lavender
- 1/2 yard lavender
- 3/4 yard gray
- 1 yard light blue
- 1 1/4 yards medium blue
- 3/4 yard dark blue
- 3/4 yard light rose
- 1/2 yard light pink
- 1/4 yard medium rose
- 4 3/4 yards white solid

PIECING INSTRUCTIONS
1. Cut four corner squares, each 22" (add seams), from white solid.

2. Cut four Z fill-in triangles (add seams).

3. Piece eight large diamonds. Each diamond is made by sewing #2 and #2 reversed to each side of #1.

4. Sew remaining star sections together by sewing B diamonds in strips and piecing to the center diamond previously constructed.

5. When eight large diamond sections are completed, sew together to make a star.

6. Set in corner squares and Z fill-in triangles to complete the top.

7. Add borders as desired.

Royal Star of DELAWARE

See DELAWARE photo on page 14

ADD
COMMON PIECES B & Z
(REFER TO PAGES 59–61)

B Cut 48 lighter lavender, 48 gray, 64 light blue, 104 medium blue, 40 dark blue, 16 light rose, 8 light pink & 56 medium rose

Z Cut 4 white solid

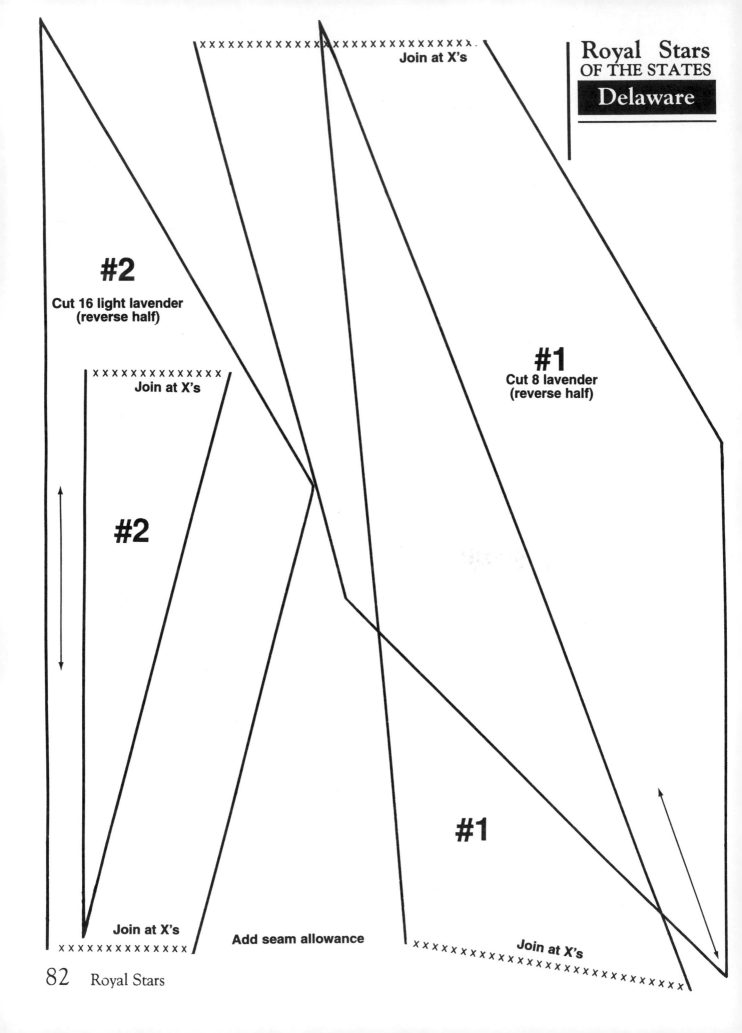

Join at X's

#2
**Cut 16 light lavender
(reverse half)**

Join at X's

#2

#1
**Cut 8 lavender
(reverse half)**

#1

Join at X's

Add seam allowance

Join at X's

72" x 72"

MATERIALS
- 1/3 yard each light pink, cream, and rose
- 3/4 yard peach
- 1/4 yard each pink and maroon
- 1 1/4 yards light rose
- 1 yard blue print
- 4 3/4 yards white solid

PIECING INSTRUCTIONS

1. Construct one side of a center point by sewing #17 to #16 to #15 to #14 to #13 to #12. To complete other side of point, reverse pieces when cutting. Join the two sides to make completed point. Sew 12 points. Adjust center to make flat, if necessary.

2. Appliqué the I piece over center.

3. Set all #11 pieces in place.

4. Sew the large outside points with pieces #10–#3 as for center points and join.

5. Set the large points in between #11 pieces. Set in the #2 pieces to make a circle.

6. Sew A pieces around outside to complete the top.

7. Add borders if desired.

Royal Star of FLORIDA

See FLORIDA photo on page 15

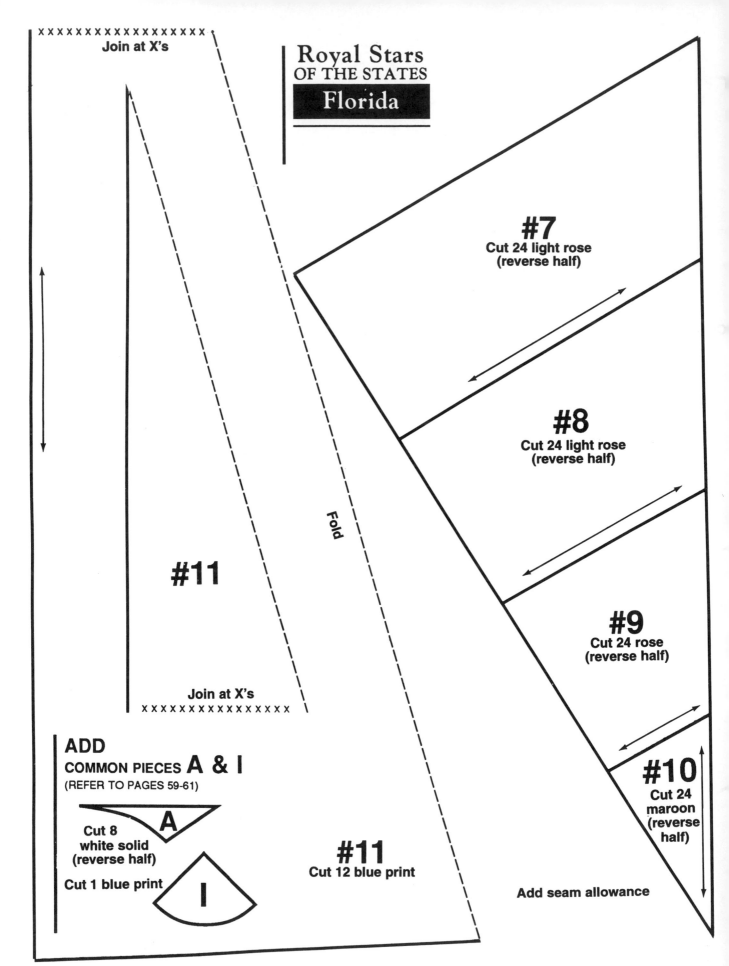

Join at X's

Royal Stars
OF THE STATES
Florida

#7
Cut 24 light rose
(reverse half)

#8
Cut 24 light rose
(reverse half)

Fold

#11

#9
Cut 24 rose
(reverse half)

Join at X's

ADD
COMMON PIECES A & I
(REFER TO PAGES 59-61)

A
Cut 8
white solid
(reverse half)

Cut 1 blue print

I

#11
Cut 12 blue print

#10
Cut 24
maroon
(reverse
half)

Add seam allowance

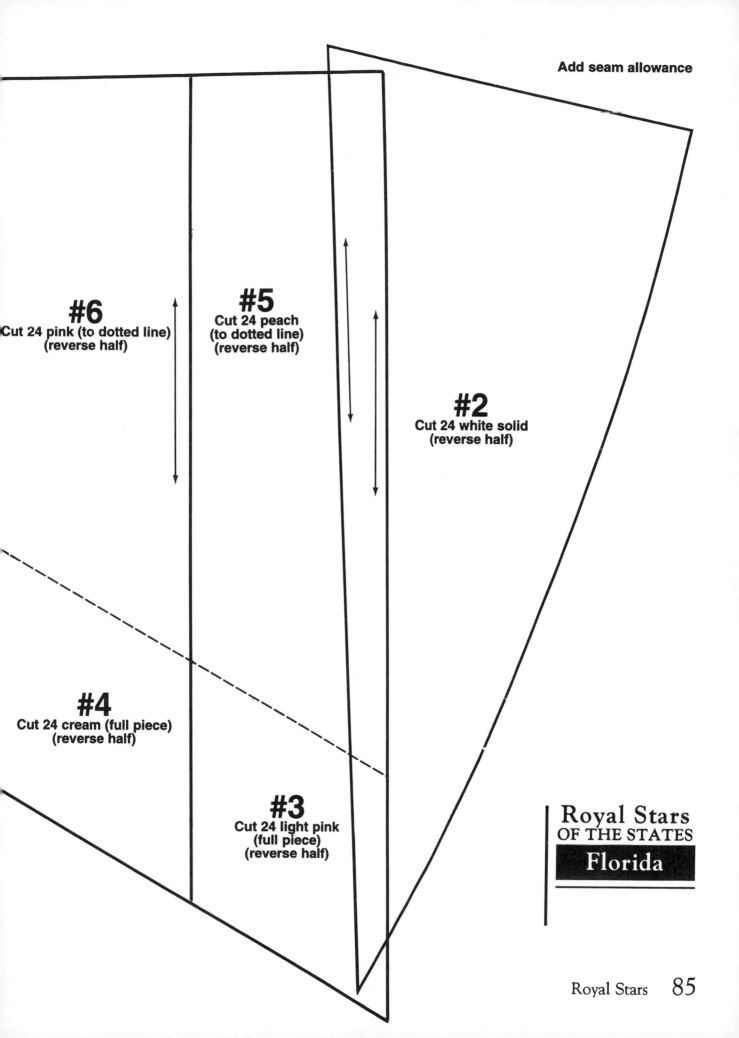

Add seam allowance

#6
Cut 24 pink (to dotted line)
(reverse half)

#5
Cut 24 peach
(to dotted line)
(reverse half)

#2
Cut 24 white solid
(reverse half)

#4
Cut 24 cream (full piece)
(reverse half)

#3
Cut 24 light pink
(full piece)
(reverse half)

Royal Stars
OF THE STATES
Florida

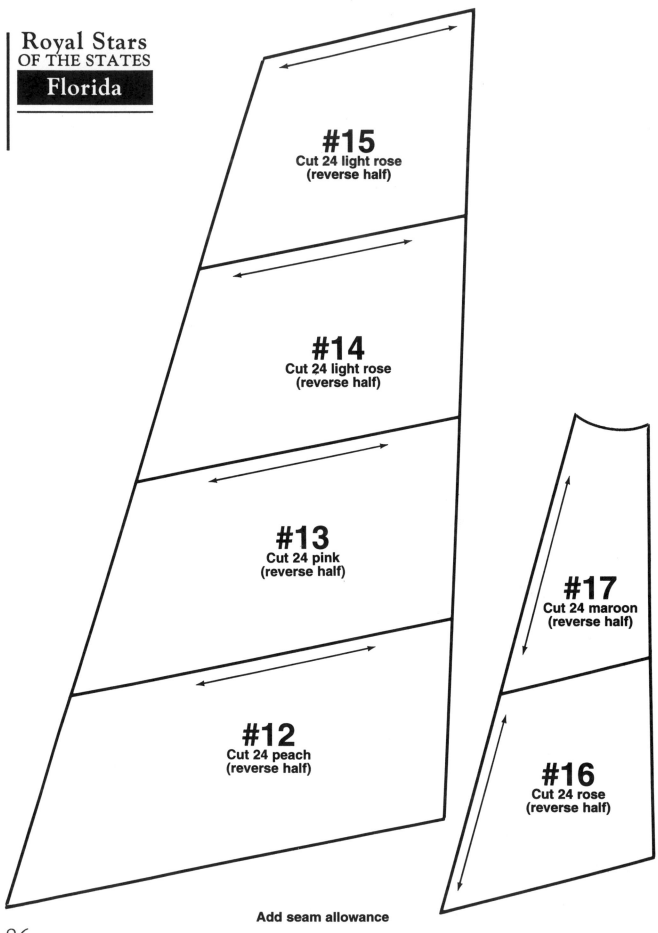

Royal Stars
OF THE STATES
Florida

#15
Cut 24 light rose
(reverse half)

#14
Cut 24 light rose
(reverse half)

#13
Cut 24 pink
(reverse half)

#17
Cut 24 maroon
(reverse half)

#12
Cut 24 peach
(reverse half)

#16
Cut 24 rose
(reverse half)

Add seam allowance

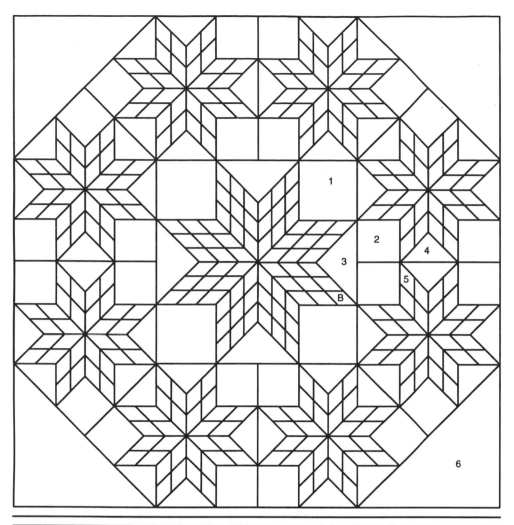

67 1/2" x 67 1/2"

MATERIALS
- 1/8 yard yellow solid
- 1/3 yard yellow print
- 1 1/4 yards brown print
- 3/4 yard gray print
- 1/2 yard rust solid
- 1/2 yard bright yellow print
- 3 yards white solid

PIECING INSTRUCTIONS

1. Cut four 8 1/4" #1 fill-in squares white solid for center of star (add seams).

2. Cut four #3 triangles white solid by cutting two 8 1/4" squares on the diagonal (add seams).

3. Cut 24 #2 fill-in squares white solid 5 3/4" (add seams).

4. Cut 40 #4 triangles white solid by cutting 20 5 3/4" (add seams) squares on the diagonal.

Royal Star of GEORGIA

See GEORGIA photo on page 16

5. Cut four large #6 corner triangles white solid by cutting two 19 3/4" squares (add seams) on the diagonal.

6. Sew small B diamonds together in rows of three. Sew three rows together to form one center star point. Repeat until there are eight star points; join the points together to complete the center star. Add fill-in triangles and squares to complete center portion.

7. Complete the smaller stars that surround the center in the same manner using piece #5.

8. When all eight star sections are complete, sew together using the fill-in squares and triangles as shown in the piecing diagram.

9. Attach these star sections to center section.

10. Sew on #6 outside corner triangles to complete the top.

11. Add borders if desired.

ADD
COMMON PIECE B
(REFER TO PAGES 59–61)

B Cut 8 yellow solid, 16 yellow print, 24 brown print, 8 gray print & 16 bright yellow print

#5
Cut 128 brown print, 64 gray print, 32 rust solid & 32 bright yellow print

72" x 72"

MATERIALS
- 1/3 yard maroon solid
- 1/3 yard blue print
- 3/4 yard lavender solid
- 2 3/4 yards maroon print
- 4 yards white solid

PIECING INSTRUCTIONS

1. Sew a #11 to each side of #12 to make one of the 12 star points. Repeat for 12 points.

2. Sew #10 to #9 to #8; repeat for reverse pieces and join. Repeat for 12 units.

3. Sew #5 to #6 to #7; repeat for reverse pieces and join. Repeat for 12 units. Sew to an #8-#9-#10 section. Sew this unit to a #11-#12 unit. Repeat for all units; join to make center circle.

4. Sew G to each short end of F for 12 sections. Sew E to each side of the remaining F pieces.

5. Join the G-F-G units and the E-F-E units around outside edge of center section to make a circle.

6. Sew A pieces to outside edges of the circle to make a square.

7. Add borders if desired.

Royal Star of HAWAII

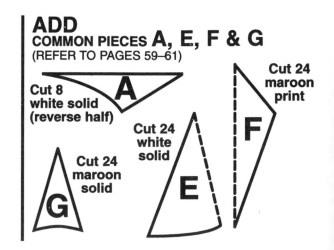

See HAWAII photo on page 17

ADD
COMMON PIECES **A, E, F & G**
(REFER TO PAGES 59–61)

A

Cut 8 white solid (reverse half)

Cut 24 maroon print

F

Cut 24 white solid

E

G

Cut 24 maroon solid

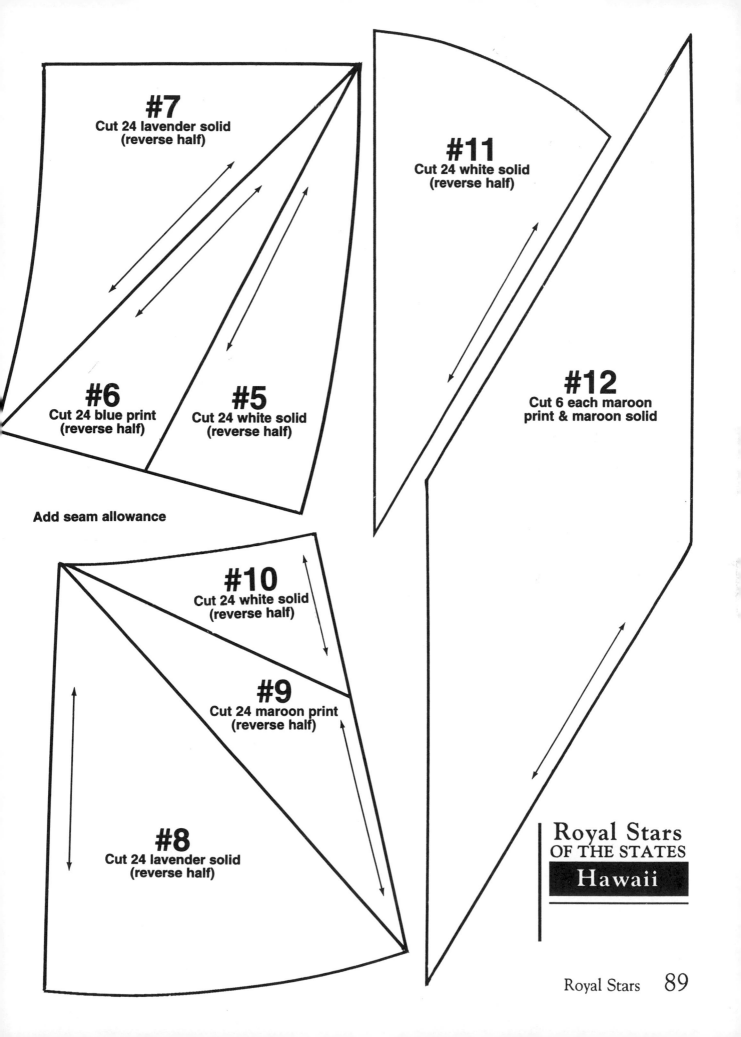

#7
Cut 24 lavender solid
(reverse half)

#11
Cut 24 white solid
(reverse half)

#6
Cut 24 blue print
(reverse half)

#5
Cut 24 white solid
(reverse half)

#12
Cut 6 each maroon
print & maroon solid

Add seam allowance

#10
Cut 24 white solid
(reverse half)

#9
Cut 24 maroon print
(reverse half)

#8
Cut 24 lavender solid
(reverse half)

Royal Stars
OF THE STATES
Hawaii

79" x 79"

MATERIALS
- 1/3 yard royal blue solid
- 1/2 yard light blue solid
- 1/2 yard pale pink solid
- 1/2 yard dark pink solid
- 3/4 yard light pink solid
- 1 1/2 yards blue print
- 4 yards white solid

PIECING INSTRUCTIONS

1. Sew center #1 diamond pieces together in rows of five. You will need 40 rows.

2. Sew the rows together in groups of five to complete eight star sections.

3. Sew the eight completed star sections together to make the center star.

4. Complete the eight outside star sections in the same manner using piece B.

5. Sew a #4 (see illustration to enlarge piece #4) to each side of an outer diamond point. Set in piece #2 (see piecing diagram).

6. Set in piece #3 to complete the circle; repeat for eight units.

7. Set these units between star points to make a circle.

8. Add piece A to make the top square.

9. Add borders if desired.

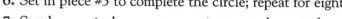

Royal Star of IDAHO

**See IDAHO
photo on page 18**

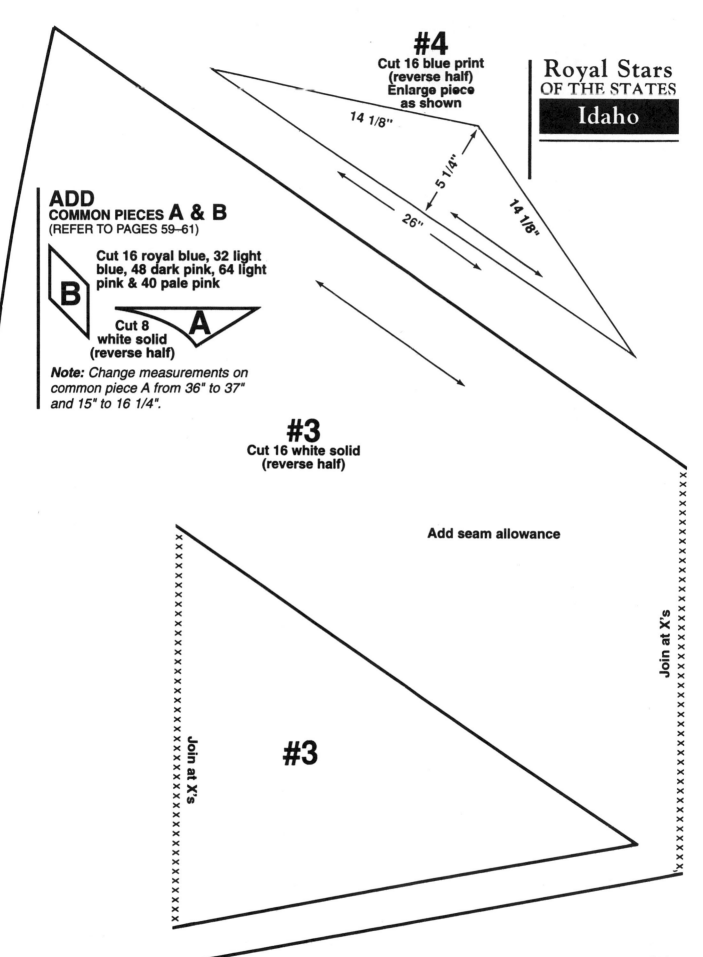

#4
Cut 16 blue print
(reverse half)
Enlarge piece
as shown

14 1/8"

5 1/4"

26"

14 1/8"

ADD
COMMON PIECES A & B
(REFER TO PAGES 59–61)

B

Cut 16 royal blue, 32 light
blue, 48 dark pink, 64 light
pink & 40 pale pink

A

Cut 8
white solid
(reverse half)

Note: Change measurements on
common piece A from 36" to 37"
and 15" to 16 1/4".

#3
Cut 16 white solid
(reverse half)

Add seam allowance

Join at X's

Join at X's

#3

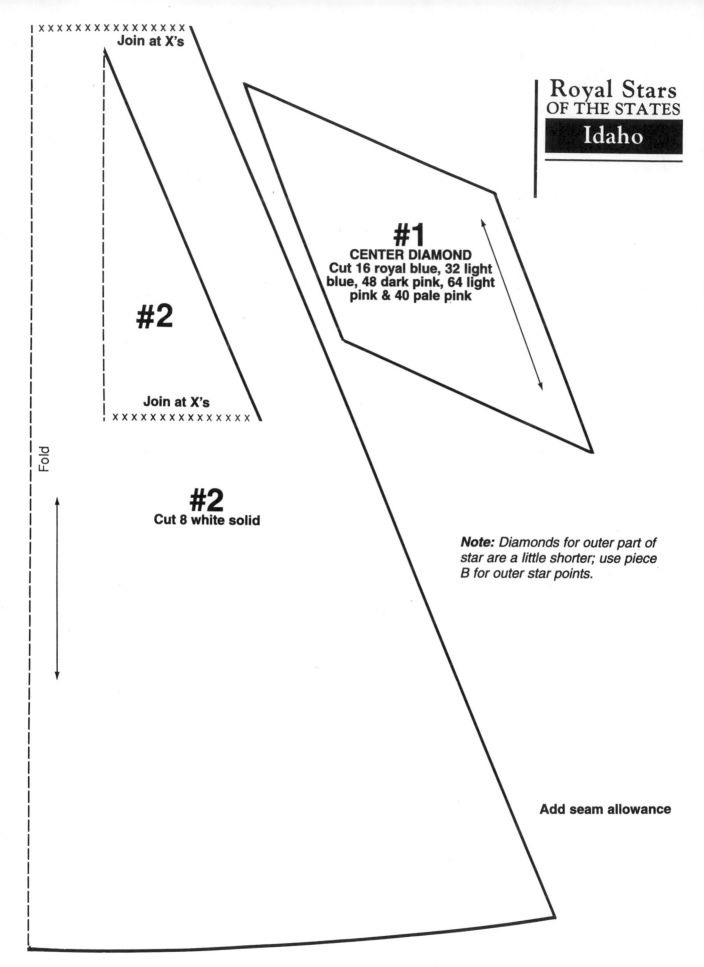

Join at X's

Join at X's

Fold

#2

#2
Cut 8 white solid

Royal Stars
OF THE STATES
Idaho

#1
CENTER DIAMOND
Cut 16 royal blue, 32 light
blue, 48 dark pink, 64 light
pink & 40 pale pink

Note: Diamonds for outer part of
star are a little shorter; use piece
B for outer star points.

Add seam allowance

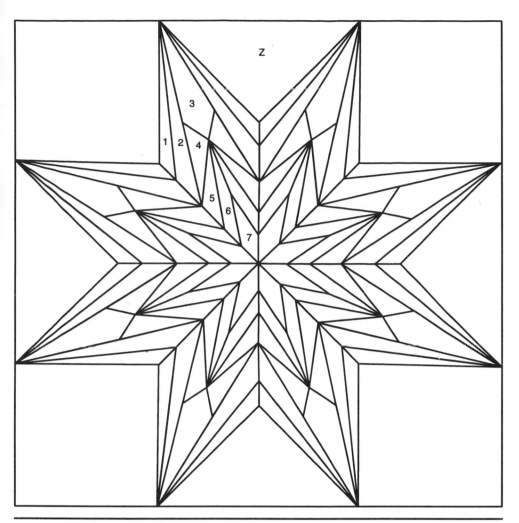

Royal Star of ILLINOIS

**See ILLINOIS photo
on page 19**

See ILLINOIS photo
on page 19

75" x 75"

MATERIALS
- 2 1/4 yards wine solid
- 2 yards wine print
- 1 1/4 yards navy print
- 3 1/2 yards white print
- 3 yards white solid

PIECING INSTRUCTIONS

Note: *Pattern pieces for this star are too large to fit on the pages. Enlarge pieces as directed.*

1. Make templates for star pieces following the instructions given with each piece. Add seam allowance to each piece before cutting.

2. To contruct star, complete eight diamond sections. To piece one section, sew #4 to #5 to #6. Repeat for reverse pieces. Join these units with #7.

3. Set #3 into #4. Sew #1 to #2; repeat for reverse pieces. Sew onto completed section to complete one star point. Repeat for eight points.

4. Join the completed star points.

5. Cut four corner squares, each 22" (add seams), from solid white. Set into corners.

6. Cut four Z fill-in triangles (add seams); set into sides to complete.

7. Add borders if desired.

Royal Stars
OF THE STATES
Illinois

Add seam allowance

Note: *Prepare templates using these measurements. Remember to add a 1/4" seam allowance to each piece when cutting and to reverse half of the pieces for right and left sides.*

#1
Cut 16 wine solid
(reverse half)

#2
Cut 16 white print
(reverse half)

#3
Cut 16 wine print

#4
Cut 14 navy print
(reverse half)

#5
Cut 16 wine print
(reverse half)

#6
Cut 16 white print
(reverse half)

#7
Cut 8 wine solid

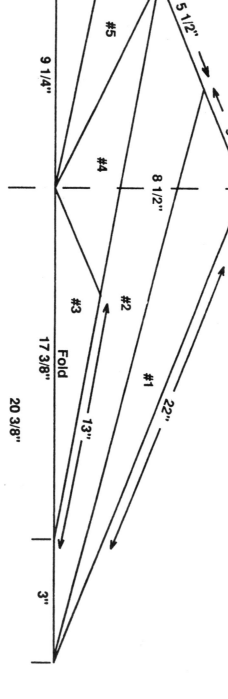

ADD
COMMON PIECE Z
(REFER TO PAGES 59-61)

Cut 4 white solid

72" x 72"

MATERIALS
- 2 1/4 yards brown print
- 1 yard brown solid
- 1/4 yard blue
- 1/2 yard light blue
- 4 1/4 yards white solid

PIECING INSTRUCTIONS

1. Sew piece #9 to #8 to #7; repeat for reverse pieces. Join to make one large center star point. Repeat for six points.

2. Appliqué H piece to center.

3. Sew #6 and #6A together; repeat for reverse pieces. Join to make a small star point. Sew #5 to each side; repeat for six units. Set these units between the large center star points to make a circle.

4. Sew #4 to each short side of F 12 times.

5. Sew E to the long sides of the remaining F pieces.

6. Sew the #4-F units and the E-F units to the center circle to make a larger circle.

7. Sew on the A pieces, matching corner seam with F points, to complete the top.

8. Add borders as desired.

Royal Star of INDIANA

See INDIANA photo on page 20

ADD
COMMON PIECES A, E, F & H
(REFER TO PAGES 59–61)

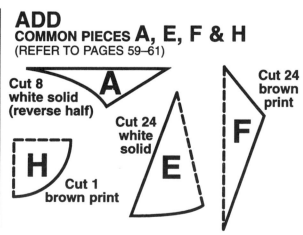

Cut 8 white solid (reverse half) **A**

Cut 24 white solid **E**

Cut 24 brown print **F**

H Cut 1 brown print

Add seam allowance

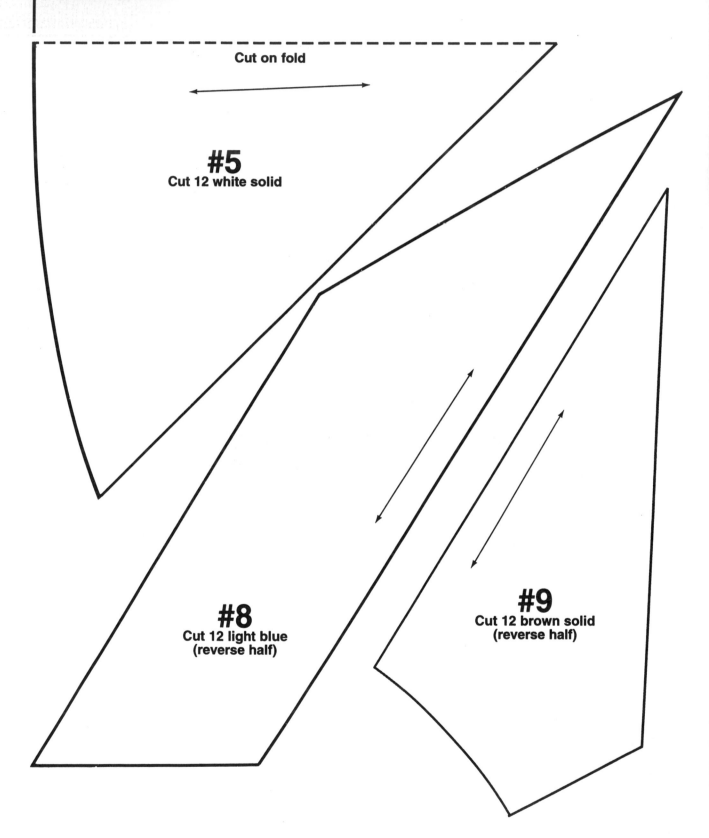

Cut on fold

#5
Cut 12 white solid

#8
**Cut 12 light blue
(reverse half)**

#9
**Cut 12 brown solid
(reverse half)**

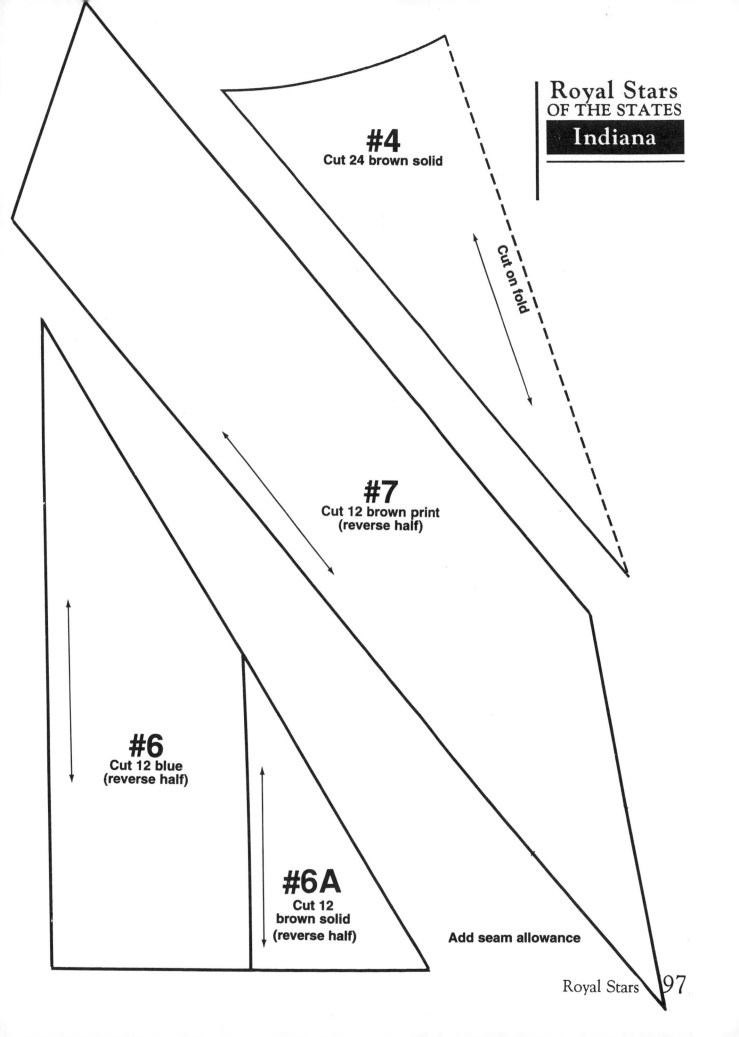

#4
Cut 24 brown solid

Cut on fold

#7
Cut 12 brown print
(reverse half)

#6
Cut 12 blue
(reverse half)

#6A
Cut 12
brown solid
(reverse half)

Add seam allowance

72" x 72"

MATERIALS
- 2/3 yard pink solid
- 1 yard wine print
- 1 yard gray print
- 1/2 yard cream solid
- 4 yards white solid

PIECING INSTRUCTIONS

1. Piece one large star point at a time to complete the quilt. Sew #7 to #6 to #5 to #4; repeat for reverse pieces. Join to make one point. Repeat for 12 points.

2. Sew #2A to #3; repeat for reverse pieces. Join along center and set onto one star point. Repeat for 12 points.

3. Set #2 pieces in between points to make a circle.

4. Sew A to corners to complete the star.

5. Add borders if desired.

Royal Star of IOWA

See IOWA photo on page 21

ADD COMMON PIECE A
(REFER TO PAGES 59–61)

A

Cut 8 white solid (reverse half)

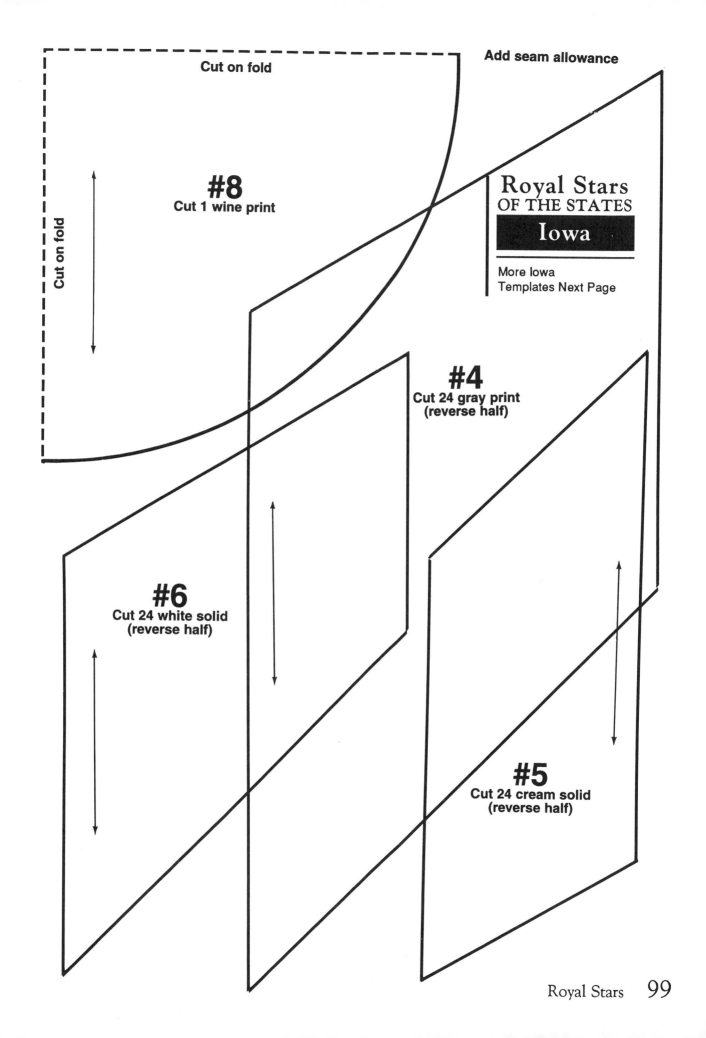

Cut on fold

Add seam allowance

Cut on fold

#8
Cut 1 wine print

Royal Stars
OF THE STATES
Iowa

More Iowa
Templates Next Page

#4
Cut 24 gray print
(reverse half)

#6
Cut 24 white solid
(reverse half)

#5
Cut 24 cream solid
(reverse half)

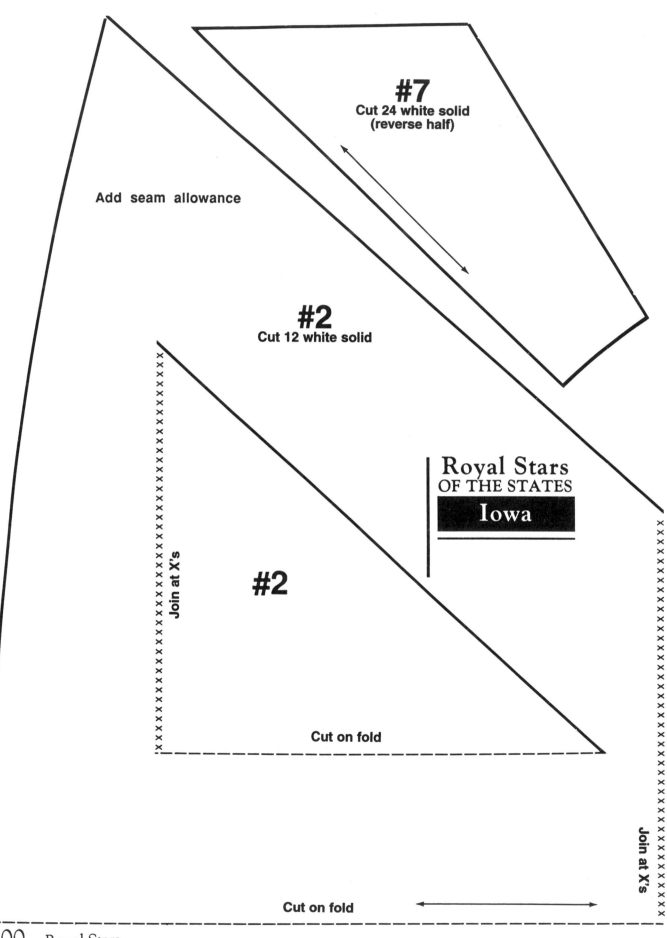

#7
Cut 24 white solid
(reverse half)

Add seam allowance

#2
Cut 12 white solid

Royal Stars
OF THE STATES
Iowa

Join at X's

#2

Cut on fold

Cut on fold

Join at X's

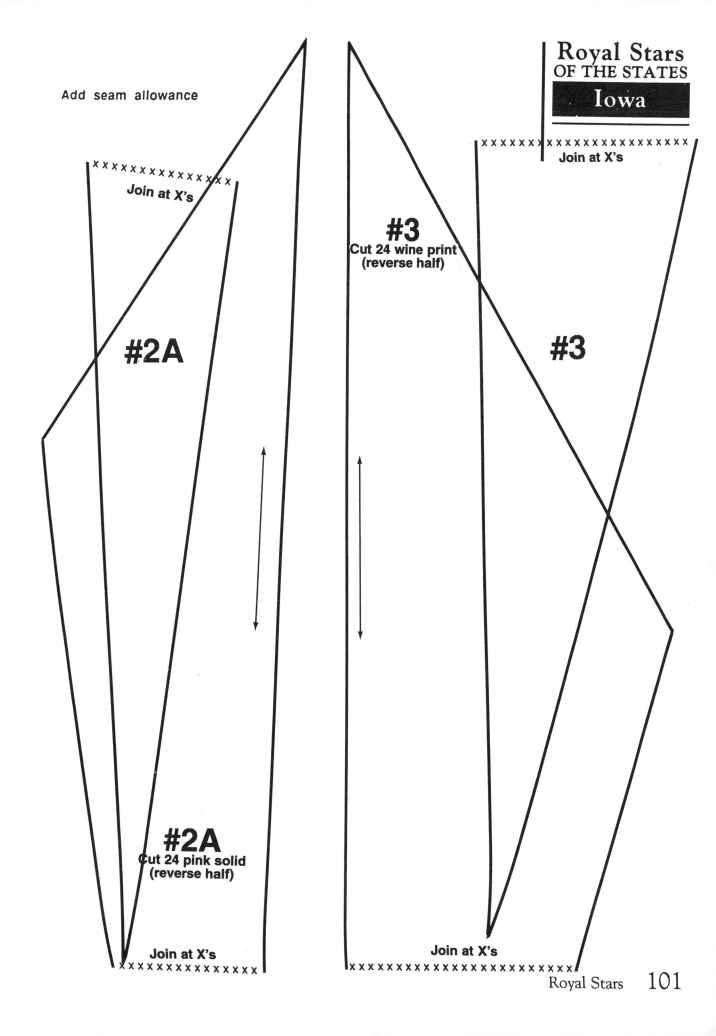

Add seam allowance

Royal Stars
OF THE STATES
Iowa

Join at X's

#2A

Join at X's

#3
Cut 24 wine print
(reverse half)

#3

#2A
Cut 24 pink solid
(reverse half)

Join at X's

Join at X's

66" x 66"

MATERIALS
- 3/4 yard blue solid #1
- 3/4 yard blue solid #2
- 1/2 yard blue solid #3
- 1/2 yard blue solid #4
- 1/4 yard blue solid #5
- 1/2 yard blue print
- 3 1/2 yards white solid

PIECING INSTRUCTIONS

Note: The quilt shown on page 22 varies from the pattern given here. The B diamonds were replaced with strips to make fewer seams.

1. Piece the center star with B pieces referring to page 22 for color arrangement or by choosing one of your own.

2. Set the #1 pieces into the star points.

3. Piece the outer star point sections in diagonal rows; join rows to complete one point. Repeat for eight points.

4. Piece the bar sections between the points using piece #2. Piece one side; reverse to complete second side; join the two sections. Set onto piece #1, sewing the star points between, in a clockwise direction.

5. Set in pieces #3, #4 and #5 to finish.

6. Add borders if desired.

Royal Star of KANSAS

See KANSAS photo on page 22

Add seam allowance

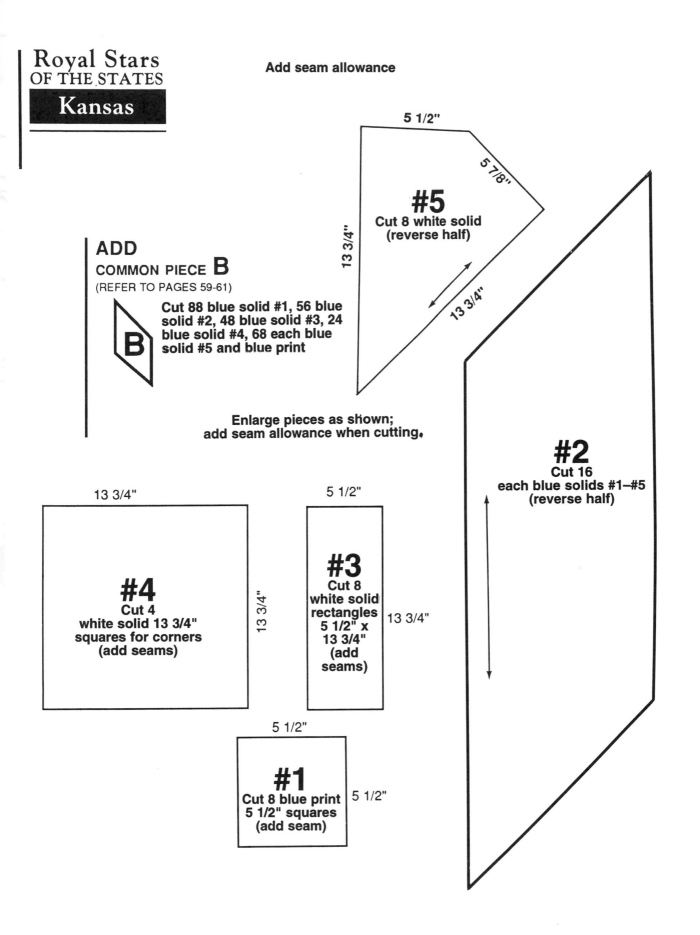

5 1/2"

5 7/8"

13 3/4"

#5
Cut 8 white solid
(reverse half)

13 3/4"

ADD

COMMON PIECE **B**

(REFER TO PAGES 59-61)

**Cut 88 blue solid #1, 56 blue
solid #2, 48 blue solid #3, 24
blue solid #4, 68 each blue
solid #5 and blue print**

B

**Enlarge pieces as shown;
add seam allowance when cutting.**

#2
Cut 16
each blue solids #1–#5
(reverse half)

13 3/4"

5 1/2"

#4
Cut 4
white solid 13 3/4"
squares for corners
(add seams)

13 3/4"

#3
Cut 8
white solid
rectangles
5 1/2" x
13 3/4"
(add
seams)

13 3/4"

5 1/2"

#1
Cut 8 blue print
5 1/2" squares
(add seam)

5 1/2"

72" x 72"

MATERIALS
- 1 1/4 yards blue solid
- 1 3/4 yards beige print
- 3/4 yard light blue print
- 3 yards white solid

PIECING INSTRUCTIONS

1. If using two fabrics for piece #3, arrange the pieces so every other piece is the same. Join all #3 pieces to make a circle.

2. Set #4 pieces in between tips.

3. Sew #2 to #1 and set in or appliqué to center.

4. Piece outer points by sewing 10 #6 triangles together 16 times (see piecing diagram and color photo for placement). Add a #7 to one end of eight sections. Sew these sections to the appropriate sides of #5 to complete each star point.

Royal Star of KENTUCKY

5. Sew the points onto inside center circle in a clockwise direction.

6. Cut four 21" squares for outside corners (add seams) from white solid.

7. Cut four Z fill-in triangles (add seams).

8. Set in corner triangles and fill-in squares to complete the top.

9. Add borders as desired.

See KENTUCKY photo on page 23

ADD COMMON PIECE Z
(REFER TO PAGES 59–61)

Z / Cut 4 white solid

Note: Change measurements on common piece Z to 21" x 21" x 30".

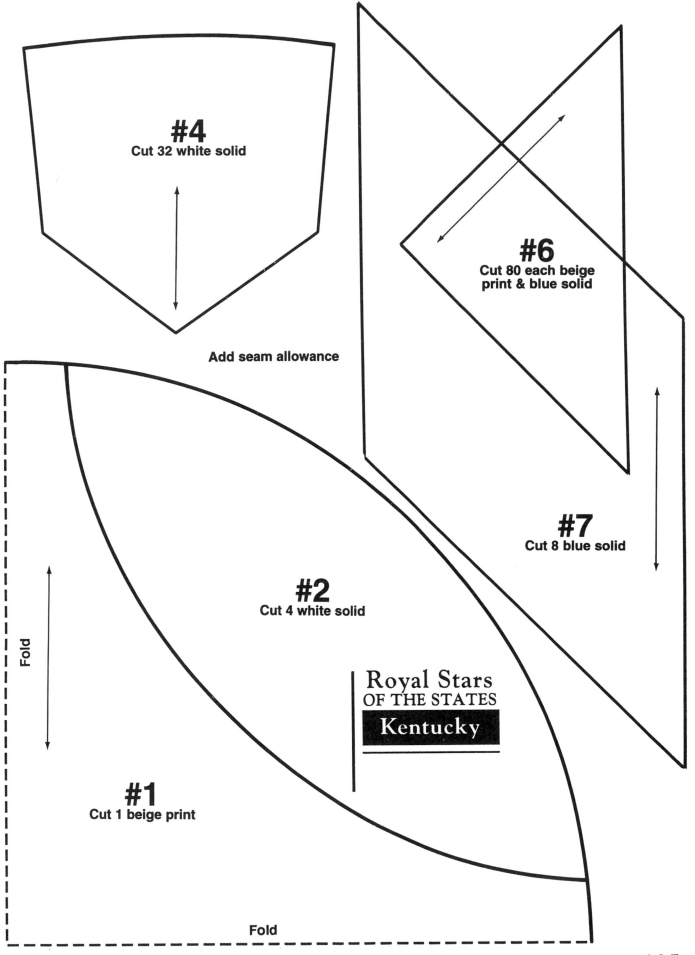

#4
Cut 32 white solid

Add seam allowance

#6
Cut 80 each beige
print & blue solid

#7
Cut 8 blue solid

Fold

#2
Cut 4 white solid

Royal Stars
OF THE STATES
Kentucky

#1
Cut 1 beige print

Fold

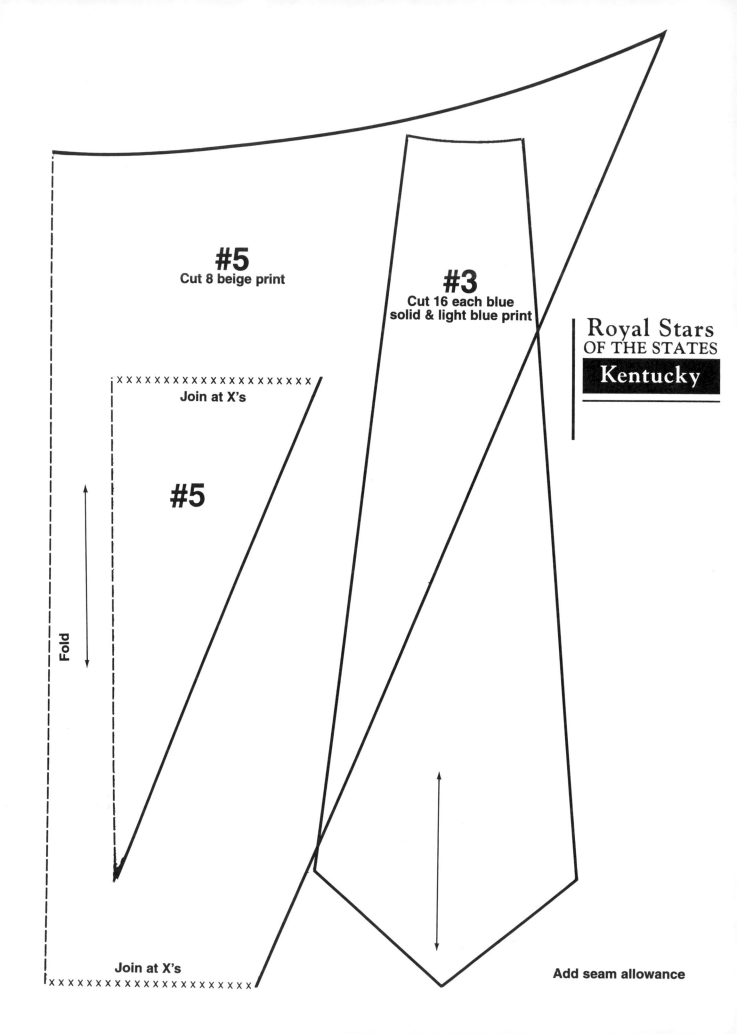

#5
Cut 8 beige print

#3
Cut 16 each blue
solid & light blue print

Royal Stars
OF THE STATES
Kentucky

x x x x x x x x x x x x x x x x x x x x x x
Join at X's

#5

Fold

Join at X's
x x x x x x x x x x x x x x x x x x x x x x

Add seam allowance

72" x 72"

MATERIALS
- 1 1/2 yards light blue print
- 1 yard dark blue print
- 3/4 yard medium blue print
- 1/4 yard light blue solid
- 1/2 yard medium blue solid
- 1/2 yard blue solid #1
- 1/3 yard blue solid #2
- 3 yards white solid

PIECING INSTRUCTIONS

1. Sew #9 to #8 to #7 to #6; repeat for reverse pieces.

2. Join the two sections to make a star point; repeat for 12 points.

3. Sew the 12 star sections together. Set in piece #5.

4. Sew #3 to #4; repeat for reverse pieces. Join the two sections to make another star point; repeat for 12 points.

5. Sew #2 and #2 reversed to sides of #3-#4 section. Repeat for 12 sections.

6. Set these sections onto the previously pieced points at piece #5.

7. Set #2A in between star points to make a circle.

8. Sew A pieces at corners to complete star top.

9. Add borders as desired.

Royal Star of LOUISIANA

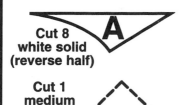

See LOUISIANA photo on page 24

ADD
COMMON PIECES A & I
(REFER TO PAGES 59–61)

A
Cut 8 white solid (reverse half)

I
Cut 1 medium blue print

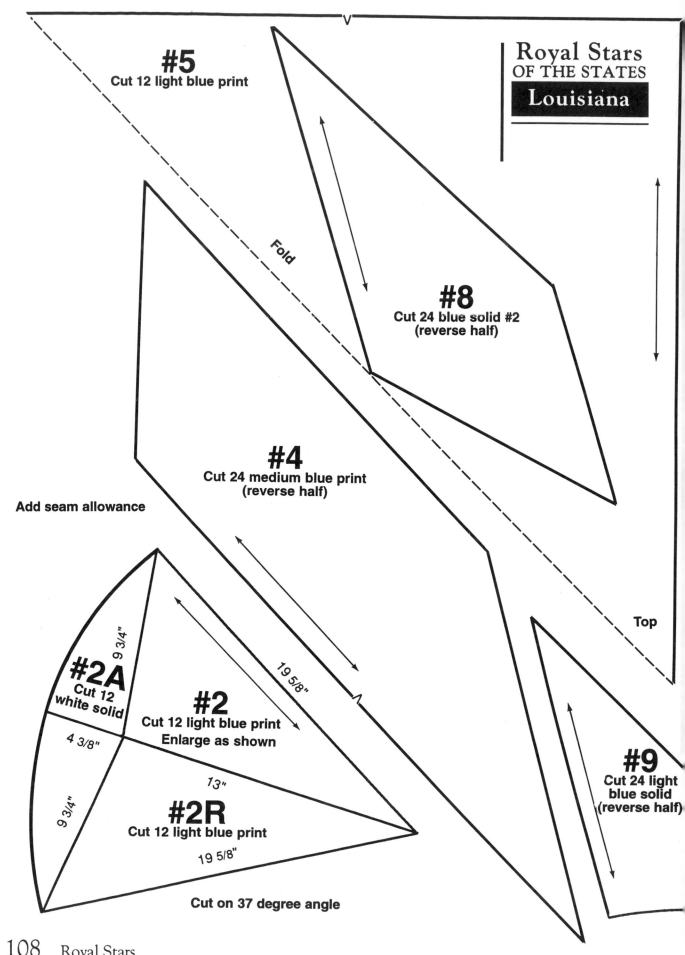

#5
Cut 12 light blue print

Royal Stars
OF THE STATES
Louisiana

Fold

#8
Cut 24 blue solid #2
(reverse half)

#4
Cut 24 medium blue print
(reverse half)

Add seam allowance

Top

#2A
Cut 12
white solid

9 3/4"

#2
Cut 12 light blue print
Enlarge as shown

19 5/8"

4 3/8"

13"

9 3/4"

#2R
Cut 12 light blue print

19 5/8"

Cut on 37 degree angle

#9
Cut 24 light
blue solid
(reverse half)

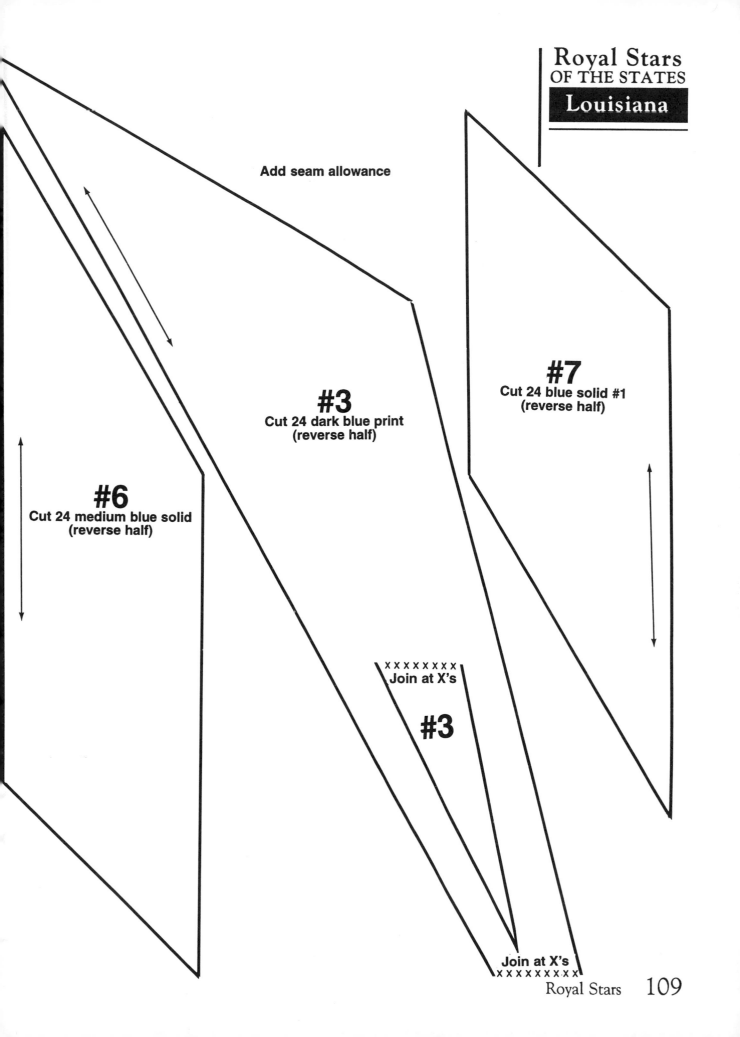

Add seam allowance

#3
Cut 24 dark blue print
(reverse half)

#7
Cut 24 blue solid #1
(reverse half)

#6
Cut 24 medium blue solid
(reverse half)

x x x x x x x x
Join at X's

#3

Join at X's
x x x x x x x x x

75" x 75"

MATERIALS
- 1 3/4 yards bright blue print
- 1 1/4 yards white print
- 3/4 yard lavender solid
- 3/4 yard brown print
- 4 1/4 yards white solid

PIECING INSTRUCTIONS

1. Piece this star in eight sections. Start in the center and work toward the outside points. Piece smaller diamonds first; sew these together to form larger diamonds; finally add the longer strips.

2. After all eight star sections are complete, piece these together to complete the star design.

3. Cut four corner squares, each 22" (add seams), from white solid.

4. Cut four Z fill-in triangles (add seams).

5. Add borders if desired.

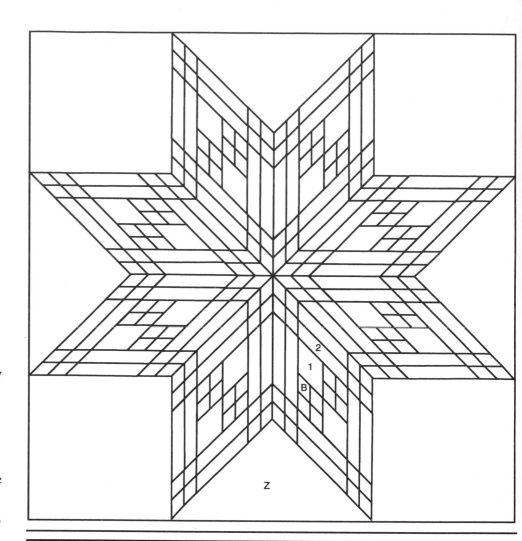

Royal Star of MAINE

See MAINE photo on page 25

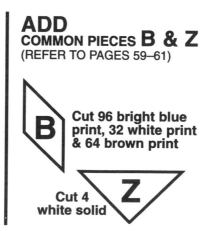

ADD
COMMON PIECES B & Z
(REFER TO PAGES 59–61)

B Cut 96 bright blue print, 32 white print & 64 brown print

Z Cut 4 white solid

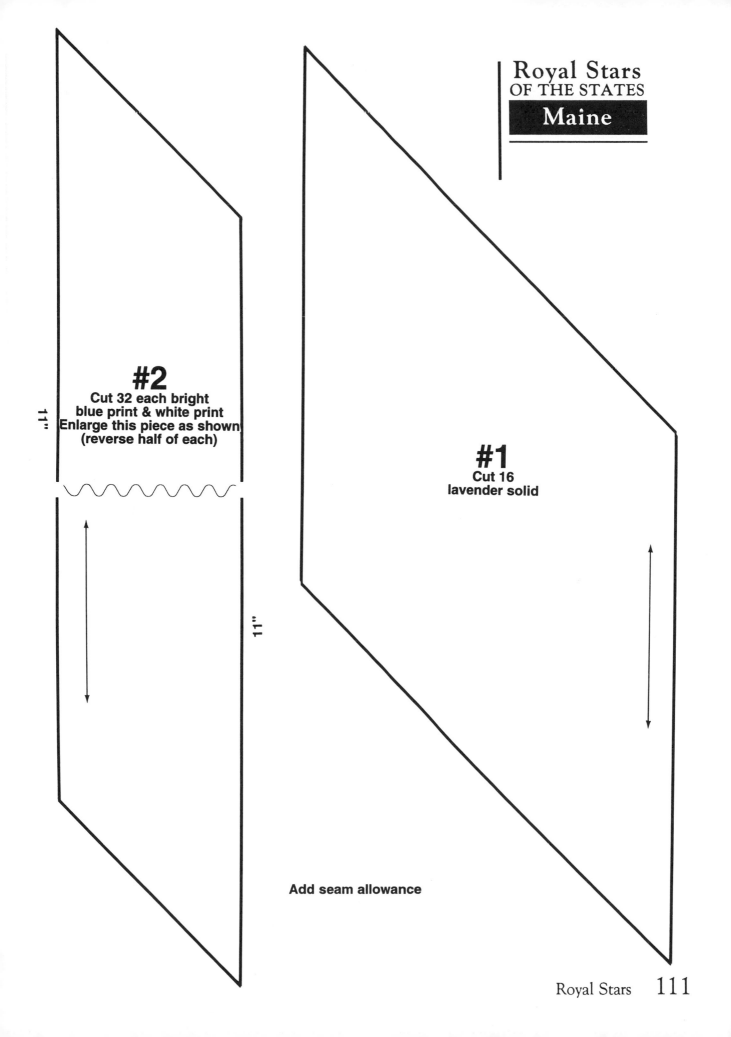

Royal Stars
OF THE STATES
Maine

#2
Cut 32 each bright
blue print & white print
Enlarge this piece as shown
(reverse half of each)

11"

11"

#1
Cut 16
lavender solid

Add seam allowance

75" x 75"

MATERIALS
- 1/2 yard each blue solid #1 and #4
- 3/4 yard each blue solid #2 and #3
- 1/2 yard each 4 different shades of pink solid
- 1 yard light pink solid #5
- 3 yards white solid

PIECING INSTRUCTIONS

1. Here is another design that is pieced in eight diamond sections. Starting in the center, sew nine B diamonds together in rows of three each referring to color photo on page 28 for color placement.

2. Sew piece #2 to #3; repeat for reverse pieces. Join at center seam; set into star points.

Royal Star of MARYLAND

3. Sew the #4, #5, #6, and #7 bar pieces together eight times; sew these to completed section.

4. Piece the outside point of the diamond in rows starting with a C and adding the B diamonds until each row is complete; join the rows to make star point. There are 28 B pieces and eight C pieces in each point.

5. Sew these completed sections to the previously pieced section to complete one large star point.

6. Sew the eight star points together to complete star design.

7. Cut four corner squares, each 22" (add seams), from white solid.

8. Cut four Z fill-in triangles (add seams).

9. Set in fill-in pieces to complete the top.

10. Add borders as desired.

See MARYLAND photo on page 26

ADD
COMMON PIECES B, C & Z
(REFER TO PAGES 59–61)

B Cut 48 blue #1, 64 blue #2, 80 blue #3, 32 blue #4, 8 pink #1, 16 pink #2, 24 pink #3, 16 pink #4 & 8 pink #5

Cut 64 blue #4 **C**

Cut 4 white solid **Z**

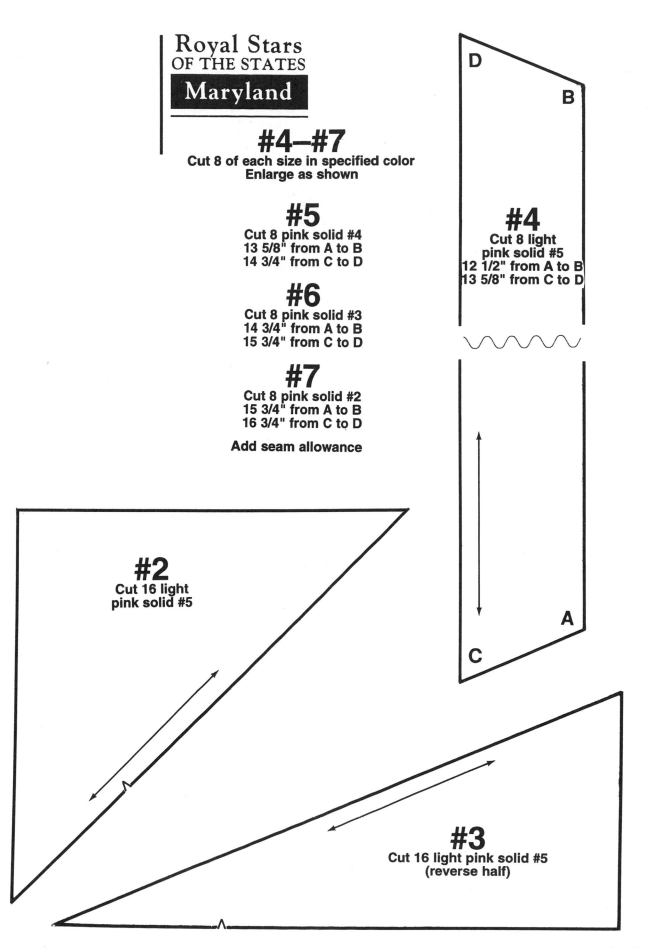

Royal Stars
OF THE STATES
Maryland

#4–#7
**Cut 8 of each size in specified color
Enlarge as shown**

#5
**Cut 8 pink solid #4
13 5/8" from A to B
14 3/4" from C to D**

#6
**Cut 8 pink solid #3
14 3/4" from A to B
15 3/4" from C to D**

#7
**Cut 8 pink solid #2
15 3/4" from A to B
16 3/4" from C to D**

Add seam allowance

D

B

#4
**Cut 8 light
pink solid #5
12 1/2" from A to B
13 5/8" from C to D**

A

C

#2
**Cut 16 light
pink solid #5**

#3
**Cut 16 light pink solid #5
(reverse half)**

66" x 66"

MATERIALS
- 1 1/4 yards light lavender
- 3/4 yard maroon solid
- 3/4 yard lavender
- 1 yard navy print
- 3 yards white solid

PIECING INSTRUCTIONS

1. Piece eight large diamonds starting in the center and working in rows to the outside referring to the diagram.

2. When eight star sections are pieced, join together to form a star design.

3. Cut four corner squares, each 19 1/4" (add seams), from white solid.

4. Cut four Z fill-in triangles (add seams).

5. Set in squares and triangles to finish quilt center. Add borders as desired.

Royal Star of MASSACHUSETTS

See MASSACHUSETTS photo on page 27

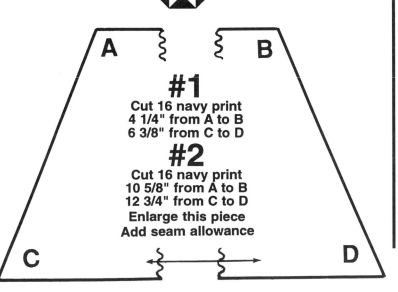

#1
Cut 16 navy print
4 1/4" from A to B
6 3/8" from C to D

#2
Cut 16 navy print
10 5/8" from A to B
12 3/4" from C to D
Enlarge this piece
Add seam allowance

ADD
COMMON PIECES B, C & Z
(REFER TO PAGES 59–61)

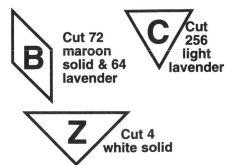

B Cut 72 maroon solid & 64 lavender

C Cut 256 light lavender

Z Cut 4 white solid

Note: Change measurement on common piece Z to 19 1/4", 19 1/4" and 27 1/2".

66 1/2" x 66 1/2"

MATERIALS
- 2 yards gray print
- 1/4 yard white print
- 1/2 yard cream solid
- 1 yard rust print
- 1 1/4 yard brown print
- 3/4 yard navy print
- 4 yards white solid

PIECING INSTRUCTIONS

1. Cut four #5 pieces following the instructions.

2. Cut eight #3 pieces following the instructions.

3. Cut 12 #4 pieces following the instructions.

4. Cut eight #6 rectangles following the instructions.

5. Piece the center star section in eight diamond units, each with 25 B diamonds.

6. Add fill-in triangles to form an octagon.

Royal Star of MICHIGAN

7. Piece half-stars using four #2 diamonds in each of the five star sections.

8. Set in small triangles between the half stars; sew to large star, setting in more triangles and rectangle pieces as you stitch.

9. Add corner triangle to complete top.

10. Add borders as desired to finish.

See MICHIGAN photo on page 28

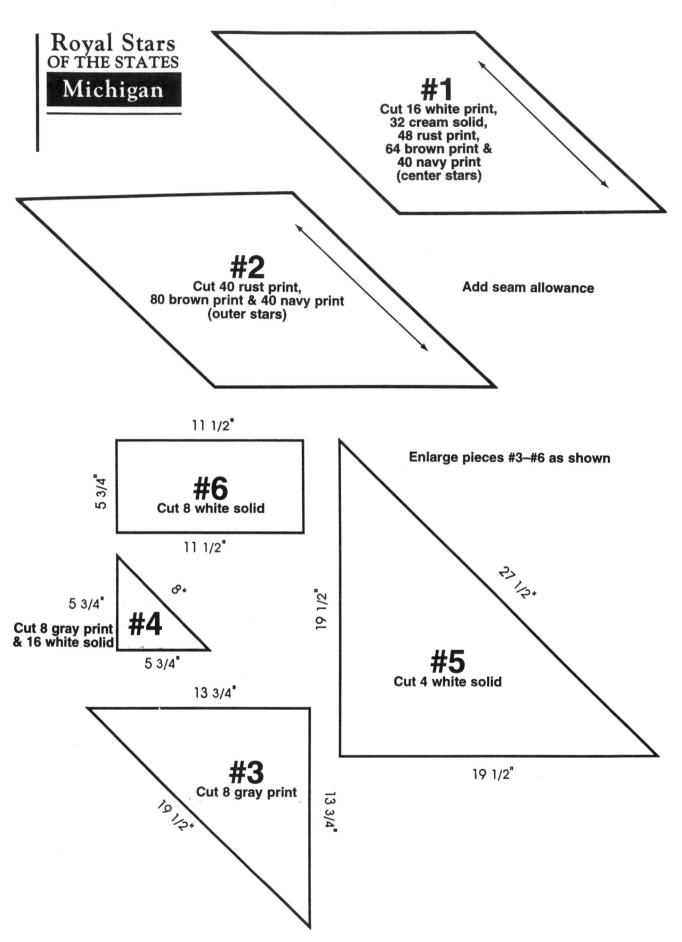

Royal Stars
OF THE STATES
Michigan

#1
Cut 16 white print,
32 cream solid,
48 rust print,
64 brown print &
40 navy print
(center stars)

#2
Cut 40 rust print,
80 brown print & 40 navy print
(outer stars)

Add seam allowance

11 1/2"

#6
Cut 8 white solid

5 3/4"

11 1/2"

Enlarge pieces #3–#6 as shown

27 1/2"

19 1/2"

#5
Cut 4 white solid

5 3/4"

8"

#4

Cut 8 gray print
& 16 white solid

5 3/4"

13 3/4"

19 1/2"

#3
Cut 8 gray print

19 1/2"

13 3/4"

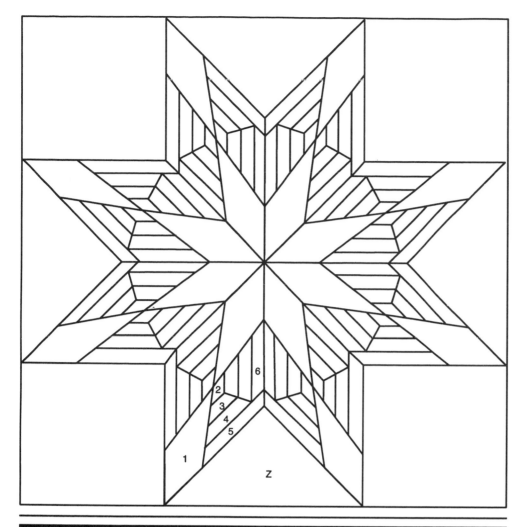

75" x 75"

MATERIALS
- 1/4 yard blue solid #1
- 1/2 yard blue solid #2
- 1/2 yard blue solid #3
- 1/2 yard blue solid #4
- 1/4 yard blue solid #5
- 1 1/2 yards dark blue print
- 3 yards white solid

PIECING INSTRUCTIONS

1. Cut four corner squares, each 22" (add seams), from white solid.

2. Cut four Z fill-in triangles (add seams).

3. Complete eight separate large diamonds using the diagram given to enlarge pieces. **Note:** *The templates for this star are too large to fit the page size. Use measurements given on drawing to make a full-size drawing. Use the drawing to create templates for each piece. Cut as directed, adding seams.*

4. Join the sections to complete the star. Set in the Z triangles and the corner squares to finish.

5. Add borders if desired.

Royal Star of MINNESOTA

See MINNESOTA photo on page 29

See MINNESOTA photo on page 29

ADD
COMMON PIECE Z
(REFER TO PAGES 59–61)

Cut 4 white solid

Z

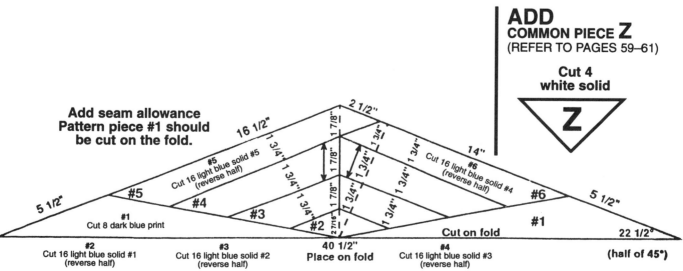

Add seam allowance
Pattern piece #1 should be cut on the fold.

5 1/2"

16 1/2"

2 1/2"

1 7/8"

1 3/4"

14"

5 1/2"

#5
Cut 16 light blue solid #5
(reverse half)

#6
Cut 16 light blue solid #4
(reverse half)

#5

#4

#3

#2

1 7/8"
1 7/8"
1 3/4"
1 3/4"
1 3/4"
1 3/4"
1 3/4"
1 3/4"
2 7/16"

#6

#1
Cut 8 dark blue print

Cut on fold

22 1/2"

#2
Cut 16 light blue solid #1
(reverse half)

#3
Cut 16 light blue solid #2
(reverse half)

40 1/2"
Place on fold

#4
Cut 16 light blue solid #3
(reverse half)

(half of 45°)

75" x 75"

MATERIALS
- 1 yard light blue solid
- 1 1/4 yards dark blue solid
- 3/4 yard blue print
- 1 yard navy solid
- 1/2 yard white print
- 1/3 yard rust print
- 1/2 yard light rust print
- 3 yards white solid

PIECING INSTRUCTIONS

1. Piece eight star shapes using piece B.

2. Starting in the center, sew #1 to #2 to #3 to #4 to #5. Repeat for reverse side and seam together down the center.

3. Sew #6 to B to #6 and insert.

4. Sew strips #8, #9 and #10 together and join to one side of piece #7.

5. Repeat for other side #8–#10 but add diamond shape made with piece B to the end before adding to #7.

6. Sew this unit to first unit pieced to complete one diamond shape. (See Figure #1.)

7. Sew the eight star points onto center to complete star design.

8. Cut four corner squares, each 22" (add seams), from white solid.

9. Cut four Z fill-in triangles (add seams).

10. Set fill-in pieces in place to complete the top.

11. Add borders if desired.

Royal Star of MISSISSIPPI

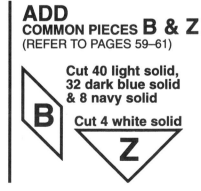

See MISSISSIPPI photo on page 30

ADD COMMON PIECES B & Z
(REFER TO PAGES 59–61)

B Cut 40 light solid, 32 dark blue solid & 8 navy solid

Z Cut 4 white solid

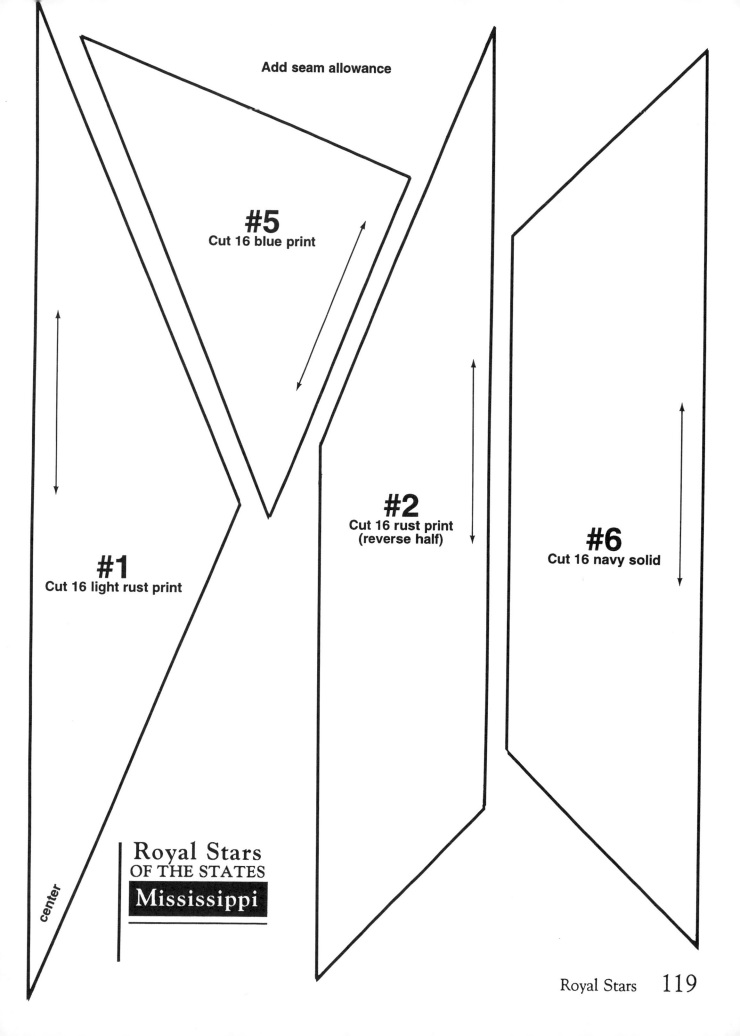

Add seam allowance

#5
Cut 16 blue print

#2
Cut 16 rust print
(reverse half)

#1
Cut 16 light rust print

#6
Cut 16 navy solid

center

Royal Stars
OF THE STATES
Mississippi

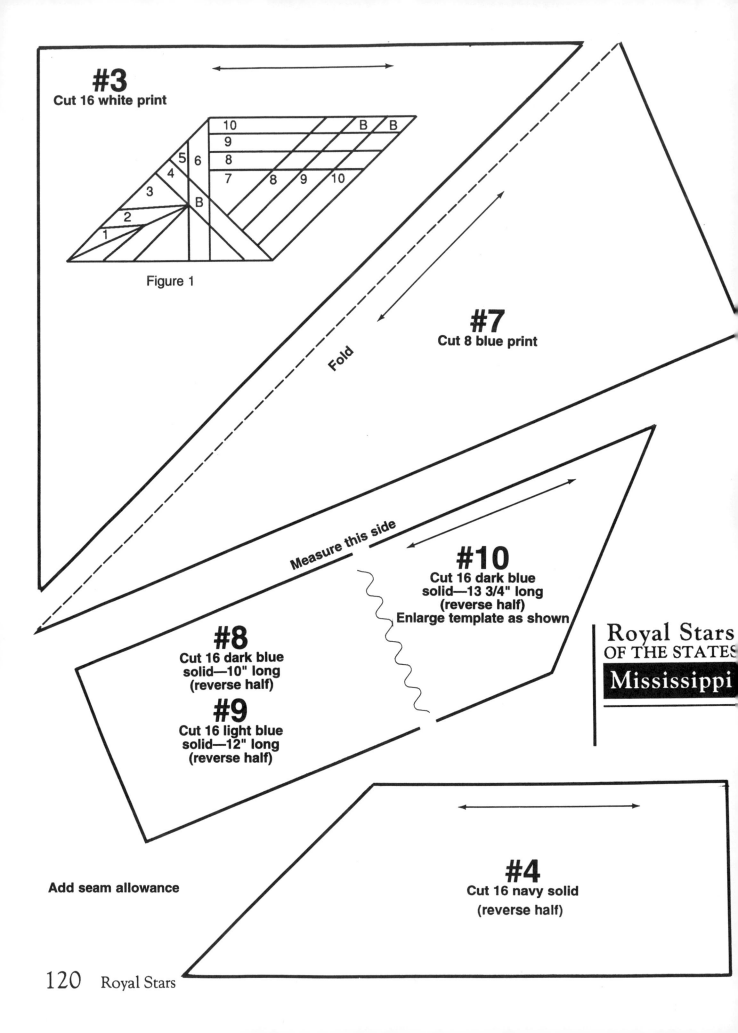

#3
Cut 16 white print

Figure 1

#7
Cut 8 blue print

Fold

Measure this side

#10
Cut 16 dark blue
solid—13 3/4" long
(reverse half)
Enlarge template as shown

#8
Cut 16 dark blue
solid—10" long
(reverse half)

#9
Cut 16 light blue
solid—12" long
(reverse half)

Royal Stars
OF THE STATES
Mississippi

Add seam allowance

#4
Cut 16 navy solid
(reverse half)

72" x 72"

MATERIALS
- 1/3 yard brown print
- 3/4 yard beige print
- 2 yards rust print
- 4 1/2 yards white solid

PIECING INSTRUCTIONS

1. Starting in the center with piece #14, add a #13 to each side, then add #12 and #11 to each side. Keep adding pieces following numbers until you reach piece G.

2. Repeat 23 times.

3. Sew the 24 sections together to form design. Appliqué I to center.

4. Sew G to each side of 12 F pieces. Sew E to each side of the remaining F pieces. Sew these units to the center design in a clockwise direction to complete circle.

5. Sew on A pieces to complete square quilt top.

6. Borders may be added if desired.

Royal Star of MISSOURI

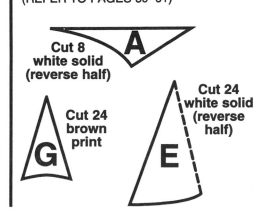

See MISSOURI photo on page 31

31

ADD
COMMON PIECES A, E, F, G & I
(REFER TO PAGES 59–61)

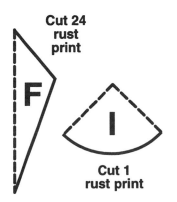

A
Cut 8 white solid (reverse half)

G
Cut 24 brown print

E
Cut 24 white solid (reverse half)

F
Cut 24 rust print

I
Cut 1 rust print

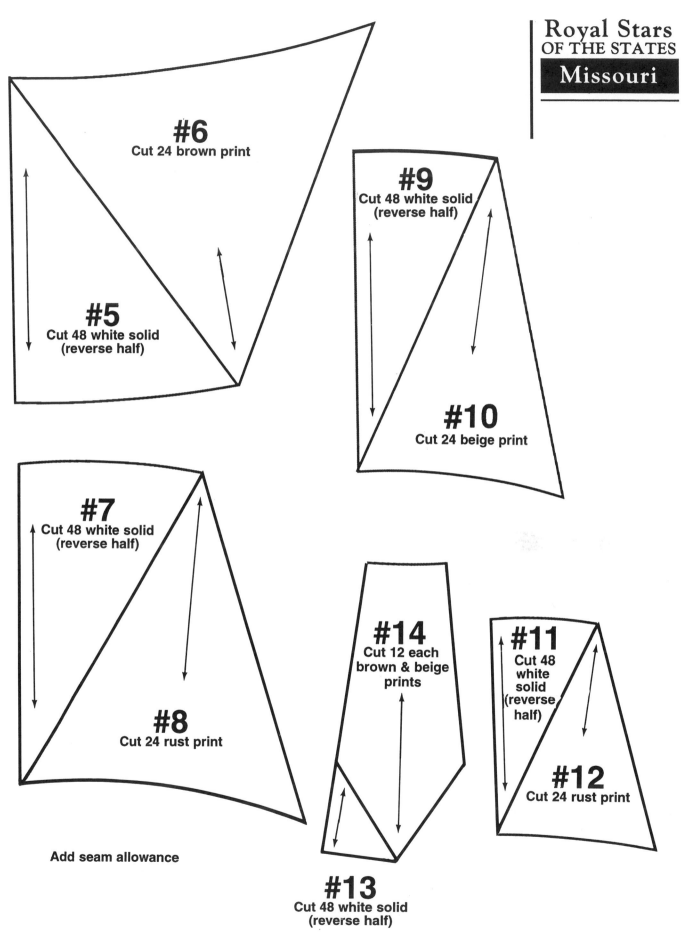

#6
Cut 24 brown print

#5
Cut 48 white solid
(reverse half)

#9
Cut 48 white solid
(reverse half)

#10
Cut 24 beige print

#7
Cut 48 white solid
(reverse half)

#8
Cut 24 rust print

#14
Cut 12 each
brown & beige
prints

#11
Cut 48
white
solid
(reverse
half)

#12
Cut 24 rust print

Add seam allowance

#13
Cut 48 white solid
(reverse half)

72" x 72"

MATERIALS
- 1 yard rose print
- 1 yard rose solid
- 1/2 yard maroon solid
- 1/2 yard blue print
- 1/3 yard light blue print
- 1/3 yard beige print
- 3 1/4 yards white solid

PIECING INSTRUCTIONS

1. Piece 12 points of each star starting in the center with piece #7. Add piece #6; repeat for reverse pieces. Set in piece #5.

2. Sew #4 and #4 reverse to the two sides of #5 and set in piece #3. Repeat for 12 star sections.

3. Join star sections to make star shape. Appliqué I over center.

4. Set in D pieces.

5. Piece 12 eight-pointed star units. Appliqué one unit to each D piece.

6. Sew A pieces to sides to finish.

7. Add borders if desired.

Royal Star of MONTANA

See MONTANA photo on page 32

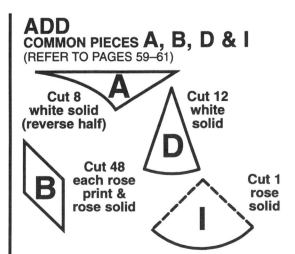

ADD
COMMON PIECES A, B, D & I
(REFER TO PAGES 59–61)

A Cut 8 white solid (reverse half)

D Cut 12 white solid

B Cut 48 each rose print & rose solid

I Cut 1 rose solid

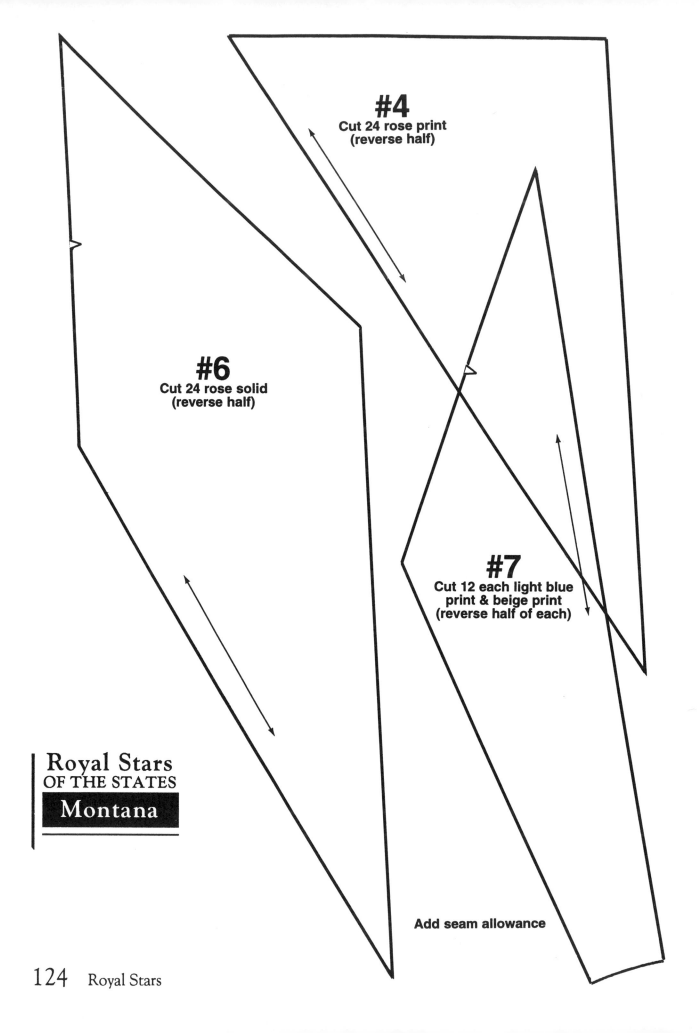

#4
Cut 24 rose print
(reverse half)

#6
Cut 24 rose solid
(reverse half)

#7
Cut 12 each light blue
print & beige print
(reverse half of each)

Royal Stars
OF THE STATES
Montana

Add seam allowance

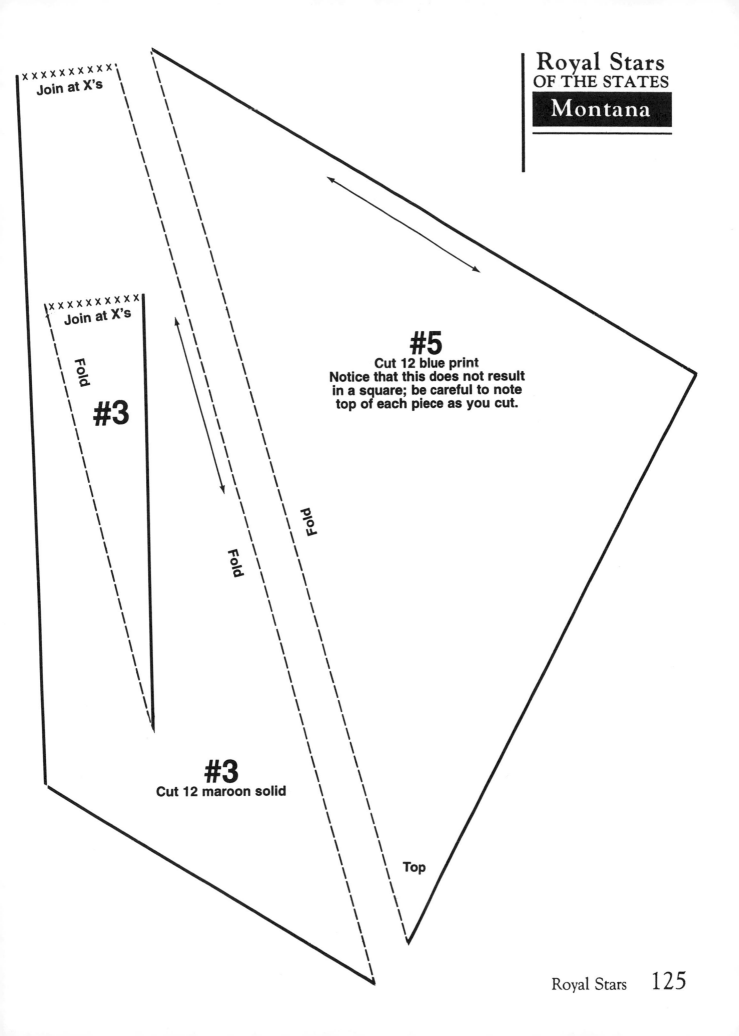

x x x x x x x x x x
Join at X's

x x x x x x x x x x x
Join at X's

Fold

#3

Fold

Fold

#5
Cut 12 blue print
Notice that this does not result
in a square; be careful to note
top of each piece as you cut.

#3
Cut 12 maroon solid

Top

72" x 72"

MATERIALS
- 1 1/2 yards rose print
- 1 yard navy print
- 1/2 yard light blue print
- 1/2 yard medium blue print
- 3 1/4 yards white solid

PIECING INSTRUCTIONS

1. Piece one star point starting with piece #7; add #6. Repeat for reverse pieces. Join at center seam.

2. Set in piece #5. Sew piece #4 to each side of #5; set in #3 to complete one star point. Repeat for 12 points.

3. Join the 12 star units to form the star design. Sew #8 pieces to the base of #7 pieces. **Note:** *Pattern for center varies slightly from photo.*

4. Set in piece D.

5. Appliqué piece I in the center.

6. Sew A pieces to outside edges to complete the top.

7. Add borders if desired.

Royal Star of NEBRASKA

See NEBRASKA photo on page 33

ADD
COMMON PIECES A, D & I
(REFER TO PAGES 59–61)

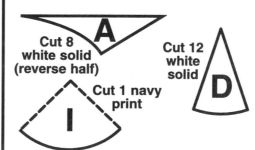

A
Cut 8 white solid (reverse half)

Cut 12 white solid
D

Cut 1 navy print
I

126   Royal Stars

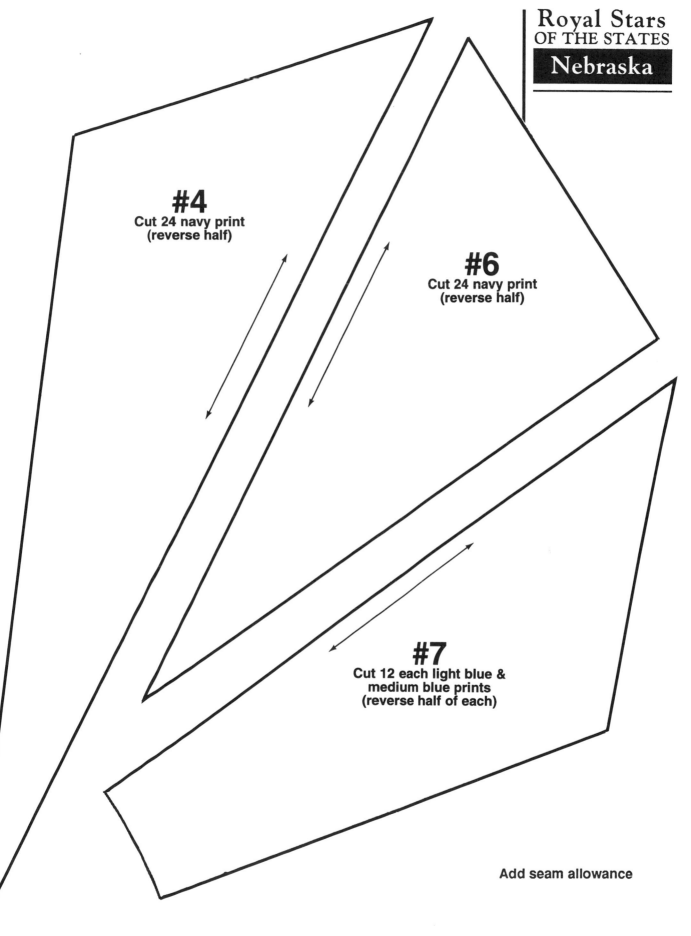

#4
Cut 24 navy print
(reverse half)

#6
Cut 24 navy print
(reverse half)

#7
Cut 12 each light blue &
medium blue prints
(reverse half of each)

Add seam allowance

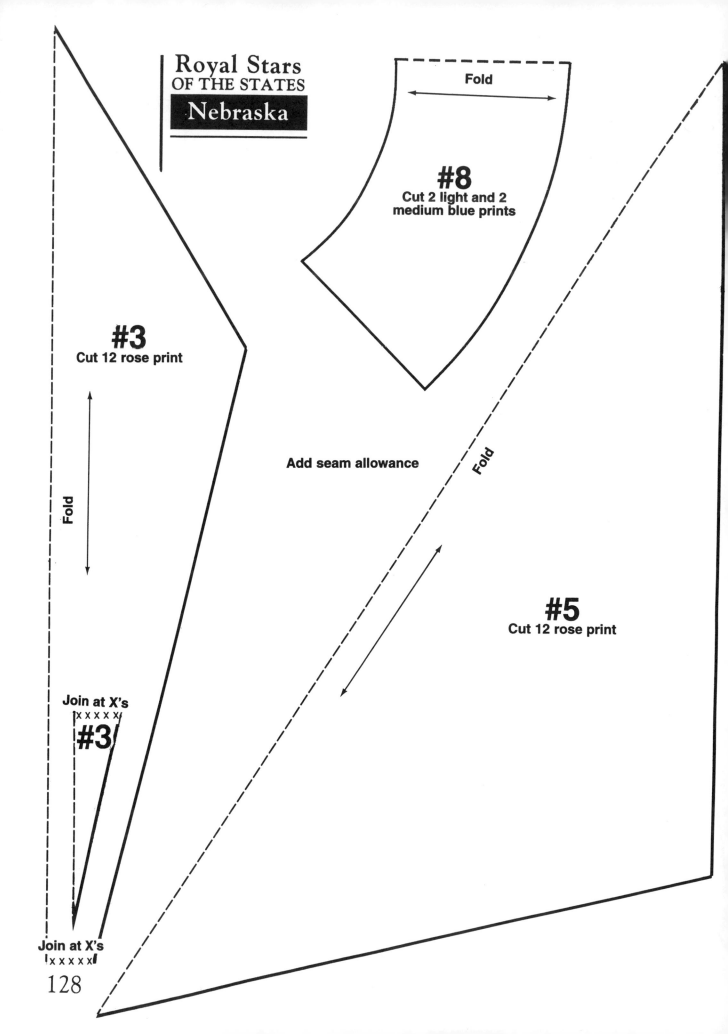

Royal Stars
OF THE STATES
Nebraska

Fold

#8
Cut 2 light and 2
medium blue prints

#3
Cut 12 rose print

Fold

Add seam allowance

Fold

#5
Cut 12 rose print

Join at X's
x x x x x

#3

Join at X's
x x x x x

128

72" x 72"

MATERIALS
- 3/4 yard light brown print
- 1 yard rose print
- 1 yard white print
- 1 yard brown print
- 4 yards white solid

PIECING INSTRUCTIONS

1. Piece together the entire center star design using the #6 diamonds.

2. Join piece #3 to #4; sew another #3 on the other side. Join to center star; repeat for eight sections.

3. Piece outer star units using the #5 diamonds.

4. Set the #5 diamond points between the points of #3.

5. Set the #2 piece in between the star points to make a circle.

6. Sew on the A corner pieces to complete the top.

7. Add borders as desired.

Royal Star of NEVADA

See NEVADA photo on page 34

ADD
COMMON PIECE A
(REFER TO PAGES 59–61)

Cut 8 white solid (reverse half)

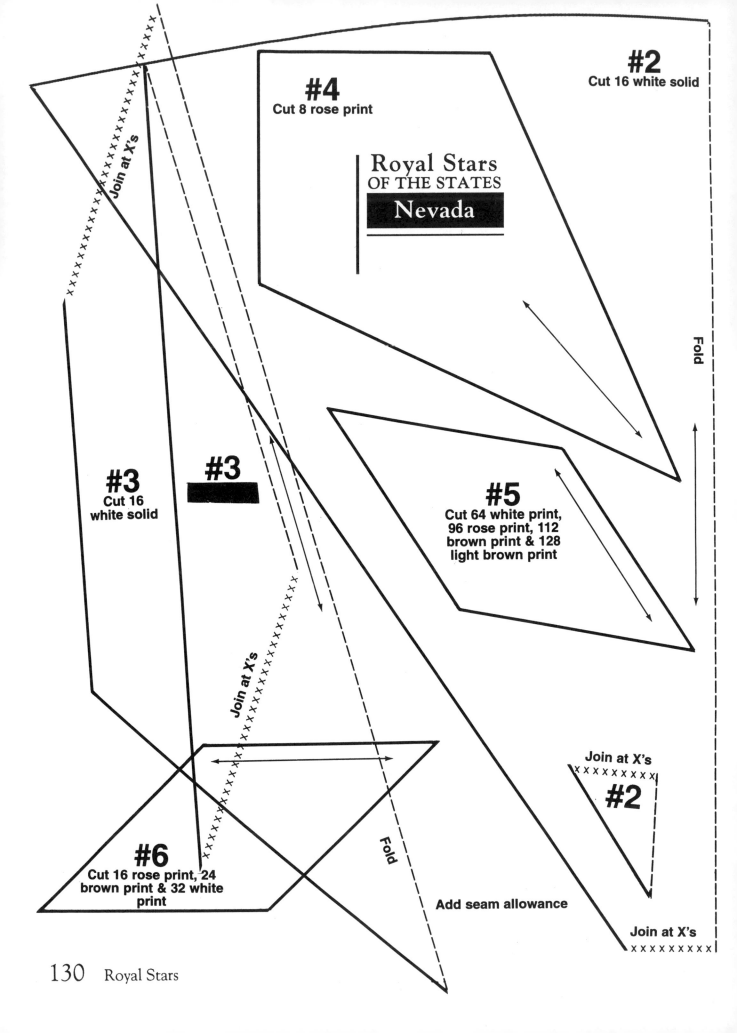

#2
Cut 16 white solid

#4
Cut 8 rose print

Royal Stars
OF THE STATES
Nevada

Fold

Join at X's

#3
Cut 16
white solid

#3

#5
Cut 64 white print,
96 rose print, 112
brown print & 128
light brown print

Join at X's

Join at X's

#2

Join at X's

Fold

#6
Cut 16 rose print, 24
brown print & 32 white
print

Add seam allowance

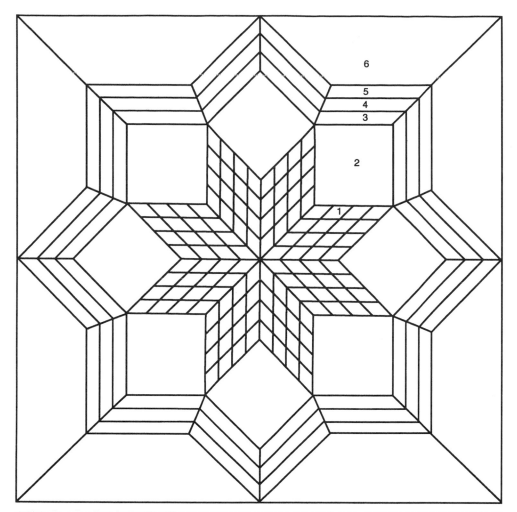

72" x 72"

MATERIALS
- 3/4 yard red print
- 1/2 yard yellow solid
- 1/2 yard yellow print
- 1 yard dark blue print
- 3 1/2 yards white solid

PIECING INSTRUCTIONS

1. Cut eight 12" squares for piece #2 from white solid (add seams).

2. Sew #1 diamonds together in the chosen color arrangement in diagonal rows; join the rows to make one star point. Repeat for eight points.

3. Join the pieced star points to make center star. Set in #2 squares.

4. Sew #3 to #4 to #5. Repeat for reverse pieces. Join at angle to make a point; set onto piece #2. Repeat for eight units.

5. Sew piece #6 all around to complete quilt top.

6. Add borders if desired.

Royal Star of NEW HAMPSHIRE

See NEW HAMPSHIRE photo on page 35

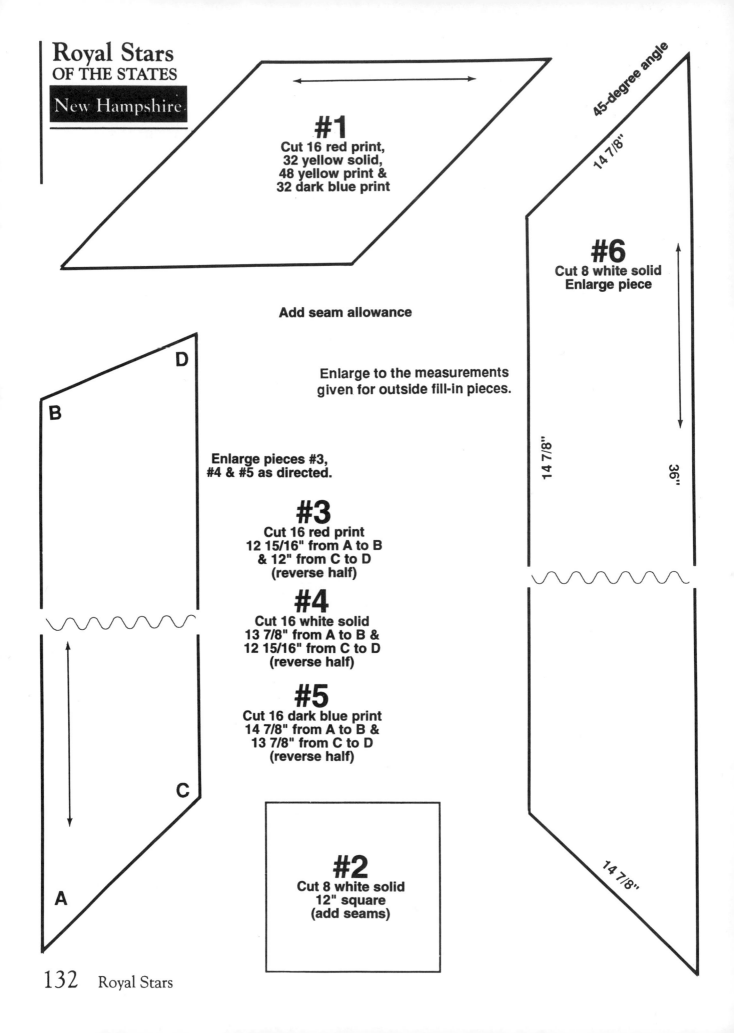

Royal Stars
OF THE STATES
New Hampshire

#1
Cut 16 red print,
32 yellow solid,
48 yellow print &
32 dark blue print

45-degree angle

14 7/8"

#6
Cut 8 white solid
Enlarge piece

14 7/8"

36"

Add seam allowance

Enlarge to the measurements
given for outside fill-in pieces.

D

B

Enlarge pieces #3,
#4 & #5 as directed.

#3
Cut 16 red print
12 15/16" from A to B
& 12" from C to D
(reverse half)

#4
Cut 16 white solid
13 7/8" from A to B &
12 15/16" from C to D
(reverse half)

#5
Cut 16 dark blue print
14 7/8" from A to B &
13 7/8" from C to D
(reverse half)

14 7/8"

C

A

#2
Cut 8 white solid
12" square
(add seams)

75" x 75"

MATERIALS
- 1 yard brown solid
- 1 yard purple print
- 1 yard peach print
- 1/2 yard rose solid
- 1/2 yard rose print
- 1/2 yard tan print
- 1/2 yard tan solid
- 1/8 yard white print
- 1/8 yard light peach print
- 3 1/2 yards white solid

PIECING INSTRUCTIONS

1. To piece each diamond section, start in the center with piece #1 and work out to #10.

2. Piece B and C in diagonal rows; join the rows to make a half star point.

3. Join the #1–#10 section to the B-C section to complete one star point. Repeat for eight points.

4. Join the star points to complete the center section.

5. Cut four corner squares, each 22" (add seams), from white solid.

6. Cut four Z fill-in triangles (add seams).

7. Insert squares and Z triangles to complete the top.

Royal Star of NEW JERSEY

***See NEW JERSEY
photo on page 36***

ADD
COMMON PIECES B, C & Z
(REFER TO PAGES 59–61)

B Cut 56 brown solid, 48 purple print, 40 peach print, 32 rose solid, 24 rose print, 16 tan print & 8 tan solid

Cut 64
white solid

C

Cut 4
white solid

Z

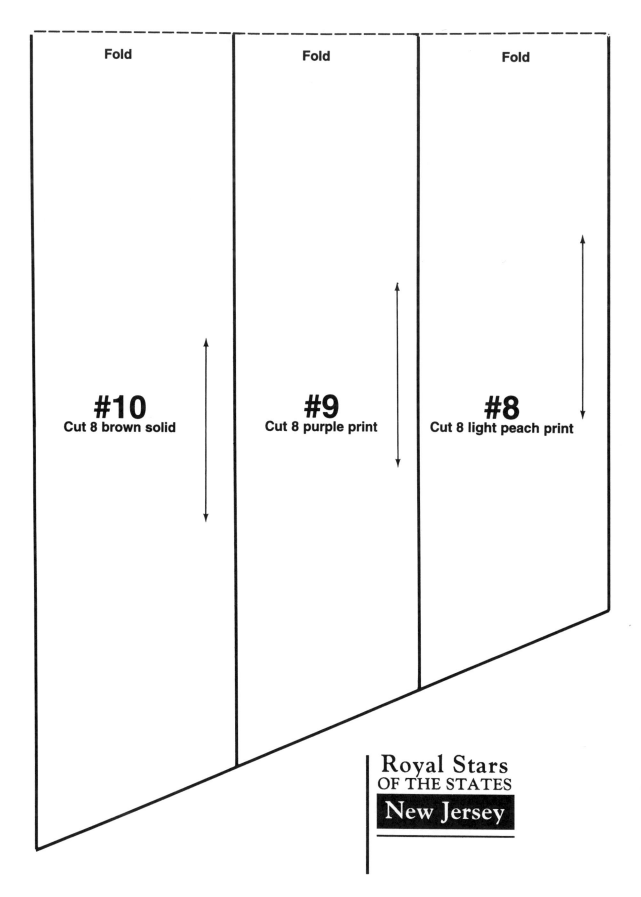

Fold

Fold

Fold

#10
Cut 8 brown solid

#9
Cut 8 purple print

#8
Cut 8 light peach print

Royal Stars
OF THE STATES
New Jersey

Add seam allowance

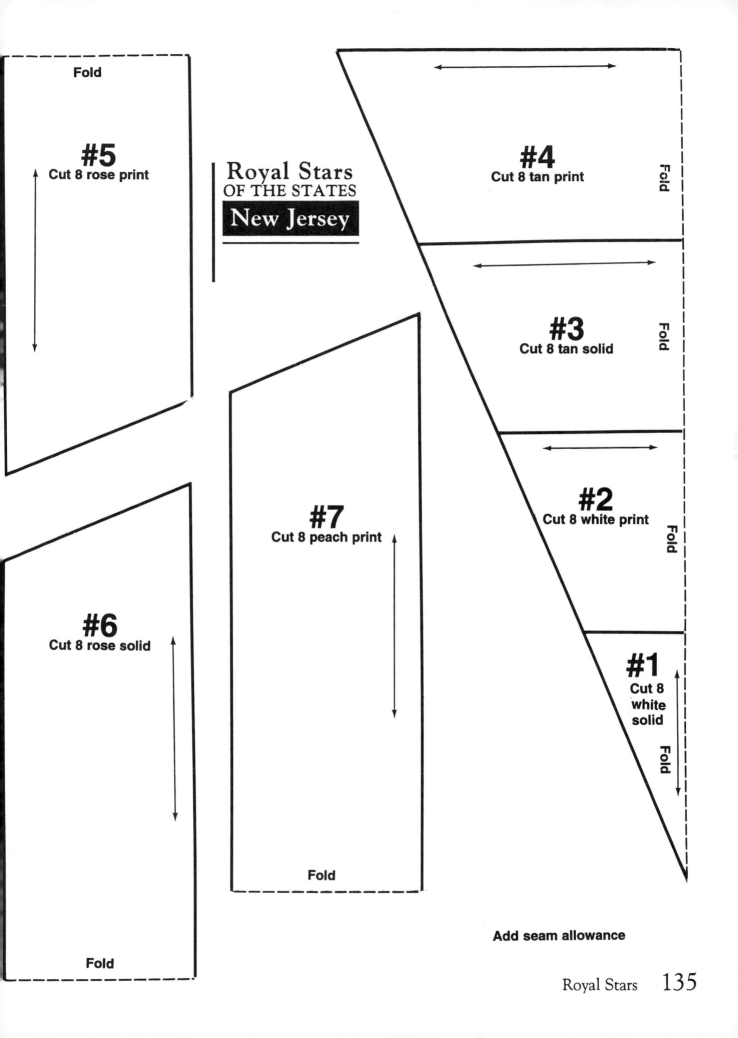

#5
Cut 8 rose print

Fold

Royal Stars
OF THE STATES
New Jersey

#4
Cut 8 tan print

Fold

#3
Cut 8 tan solid

Fold

#7
Cut 8 peach print

#2
Cut 8 white print

Fold

#6
Cut 8 rose solid

#1
Cut 8
white
solid

Fold

Fold

Fold

Add seam allowance

75" x 75"

MATERIALS
- 1 1/4 yards rust solid
- 3/4 yard rust print
- 3/4 yard beige print
- 3/4 yard royal blue print
- 3 yards white solid

PIECING INSTRUCTIONS

1. There are three ways to construct the eight large diamonds that form the star in the pattern. You may sew from the center out, or sew in either horizontal or vertical rows. Choose the method which seems easier for you.

2. Cut four corner squares, each 22" (add seams), from white solid.

3. Cut four Z fill-in triangles (add seams).

4. After star points are constructed and joined to make center design, set in large squares and Z triangles to complete the design.

5. If desired, add a border before finishing the quilt.

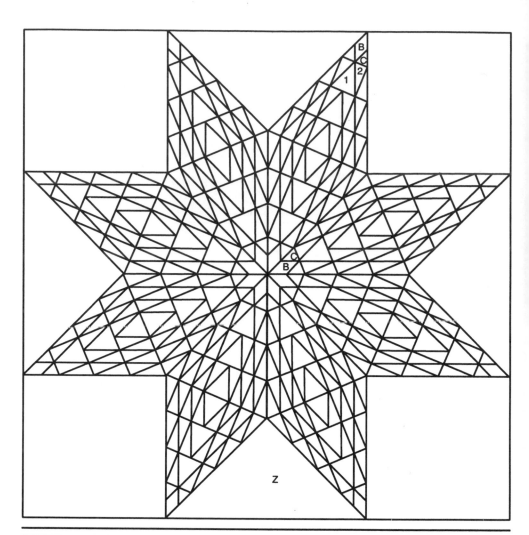

Royal Star of NEW MEXICO

See NEW MEXICO photo on page 37

ADD
COMMON PIECES B, C & Z
(REFER TO PAGES 59–61)

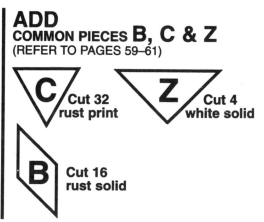

C Cut 32 rust print

Z Cut 4 white solid

B Cut 16 rust solid

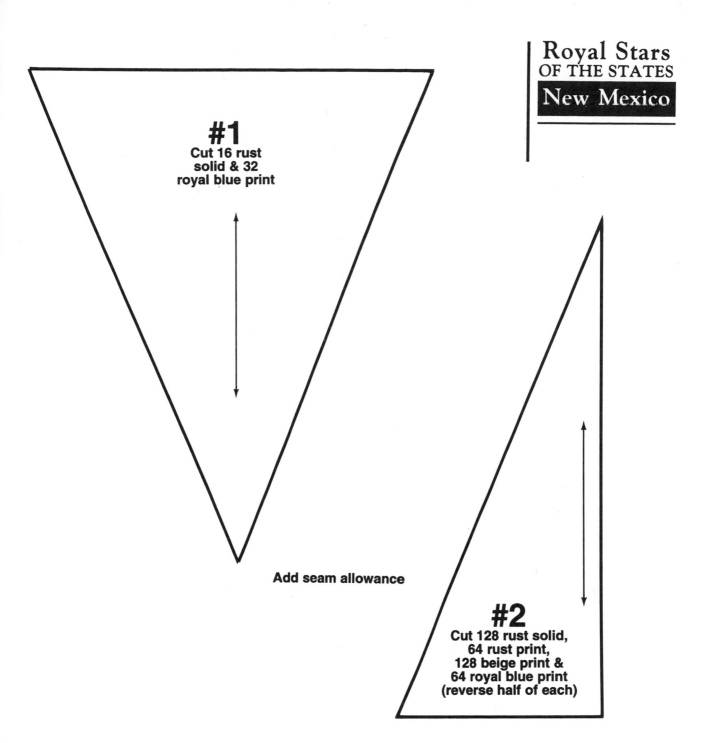

#1
**Cut 16 rust
solid & 32
royal blue print**

Add seam allowance

#2
**Cut 128 rust solid,
64 rust print,
128 beige print &
64 royal blue print
(reverse half of each)**

72" x 72"

MATERIALS
- 1/2 yard gray print
- 2 1/2 yards rust print
- 1/2 yard cream solid
- 1/3 yard beige print
- 3/4 yard pale yellow solid
- 3 yards white solid

PIECING INSTRUCTIONS

1. Sew all #11 pieces together to form a circle for center. Set in #10 pieces.

2. Appliqué H center piece in place.

3. Set on pieces #8 and #9 to form a second circle.

4. Construct third circle with #5, #6 and #7.

5. Sew two G pieces to each of 12 F pieces and set onto circle.

6. Sew E to each side of the remaining F's and set between points to make another circle.

7. Set A pieces on outside edges to make a square to finish the top.

8. Add borders to finish if desired.

Royal Star of NEW YORK

See NEW YORK photo on page 38

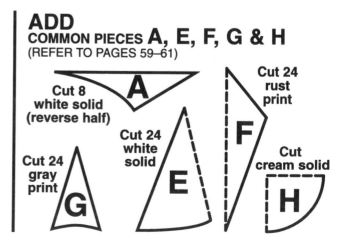

ADD COMMON PIECES A, E, F, G & H
(REFER TO PAGES 59–61)

Cut 8 white solid (reverse half) — A

Cut 24 white solid — E

Cut 24 rust print — F

Cut 24 gray print — G

Cut cream solid — H

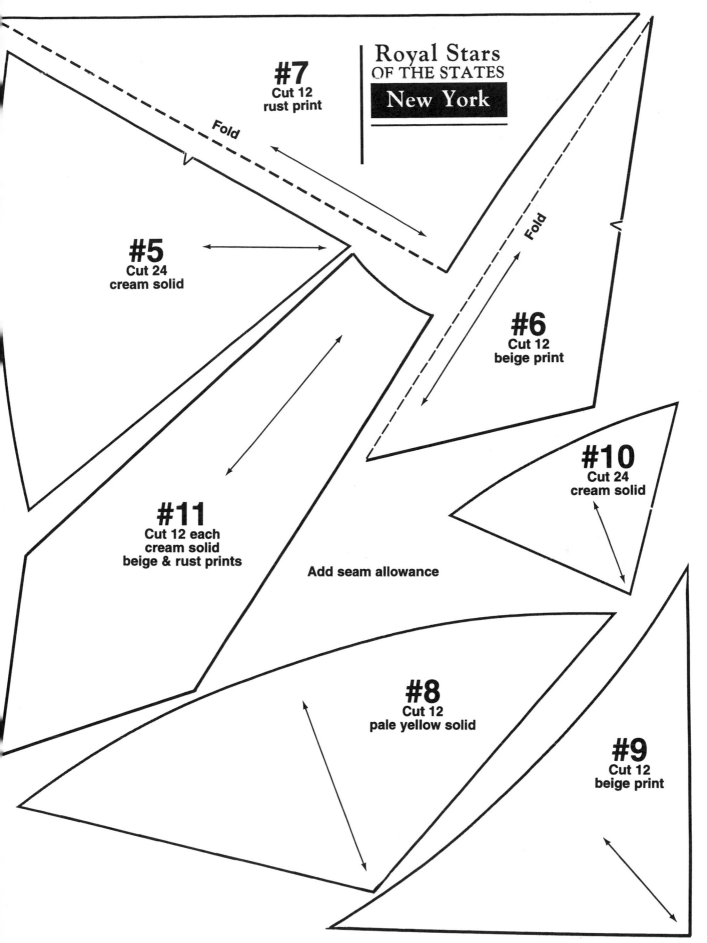

#7
Cut 12
rust print

Royal Stars
OF THE STATES
New York

Fold

Fold

#5
Cut 24
cream solid

#6
Cut 12
beige print

#10
Cut 24
cream solid

#11
Cut 12 each
cream solid
beige & rust prints

Add seam allowance

#8
Cut 12
pale yellow solid

#9
Cut 12
beige print

75" x 75"

MATERIALS
- 2 3/4 yards black solid
- 2 3/4 yards orange solid
- 7 yards white solid

PIECING INSTRUCTIONS

1. Piece eight diamond shapes separately.

2. Start at the inside and work out. It is easier if the left and right sides of each diamond are pieced first, then seamed down the center.

3. When all eight diamond sections are complete, join together to complete star design.

4. Cut four 22" squares for corner fill-in pieces (add seams), from white solid.

5. Cut four Z fill-in triangles (add seams).

6. Set in squares and Z triangles to complete the quilt top.

7. Add borders, if desired.

Royal Star of NORTH CAROLINA

See NORTH CAROLINA photo on page 39

ADD
COMMON PIECE Z
(REFER TO PAGES 59–61)

Cut 4 white solid Z

Note: This pattern has many large pieces. Since these pieces will not fit on a page, please refer to the scale drawing. Measurements given are finished size—add seam allowance when cutting.

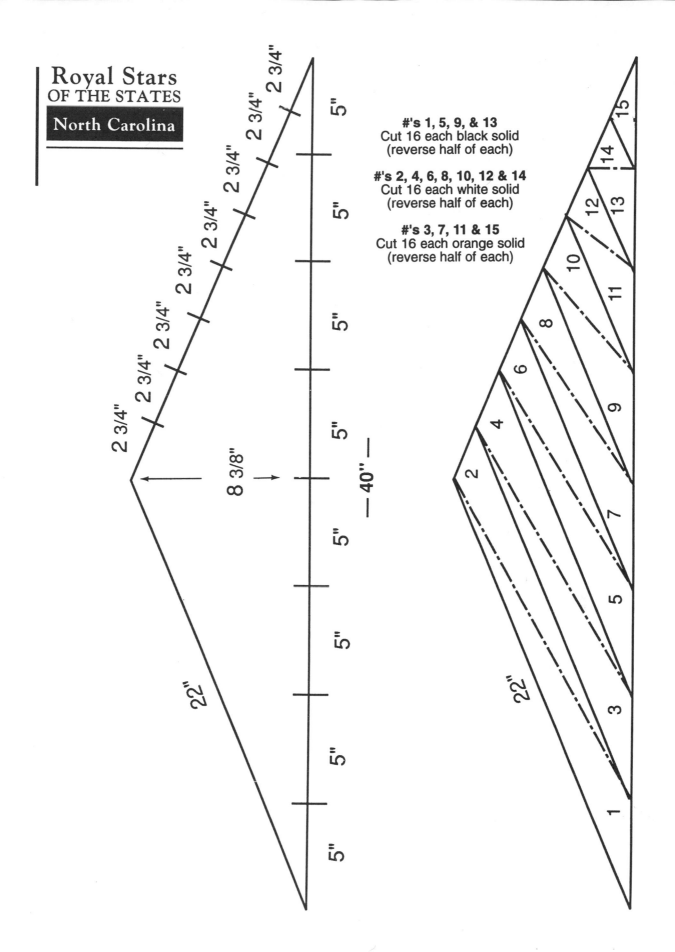

Royal Stars
OF THE STATES
North Carolina

#'s 1, 5, 9, & 13
Cut 16 each black solid
(reverse half of each)

#'s 2, 4, 6, 8, 10, 12 & 14
Cut 16 each white solid
(reverse half of each)

#'s 3, 7, 11 & 15
Cut 16 each orange solid
(reverse half of each)

5"
5"
5"
5"
5"
5"
5"
5"

2 3/4"
2 3/4"
2 3/4"
2 3/4"
2 3/4"
2 3/4"
2 3/4"
2 3/4"

8 3/8"

— 40" —

22"

22"

75" x 75"

MATERIALS
- 3/4 yard yellow print
- 1 yard rust print
- 1 yard yellow solid
- 3/4 yard blue/rust print
- 3 yards white solid

PIECING INSTRUCTIONS

1. Sew three B diamonds in three diagonal rows in your chosen color arrangement. Join the rows to complete one star point for center star; repeat for eight points.

2. Join the star points to complete center star. Sew a #1 piece to one side. Sew #2 to #1 and sew to other side. Repeat for each star point.

3. Sew three #3 strips together; repeat for reverse pieces. Sew one pieced #3 unit to one side of a star point.

4. Sew another B star point section as in step 1. Sew a #3 strip unit to one side and set onto the opposite side of the same star point. Repeat for all star points.

5. Cut four 22" corner squares (add seams) from white solid.

6. Cut four Z fill-in triangles (add seams).

7. Set in Z triangles and corner squares to complete quilt top.

8. If desired, add a border before finishing the quilt.

Royal Star of NORTH DAKOTA

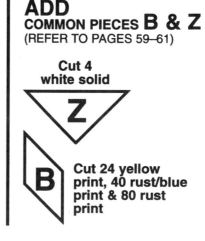

See NORTH DAKOTA photo on page 40

ADD
COMMON PIECES B & Z
(REFER TO PAGES 59–61)

Cut 4 white solid

Z

B Cut 24 yellow print, 40 rust/blue print & 80 rust print

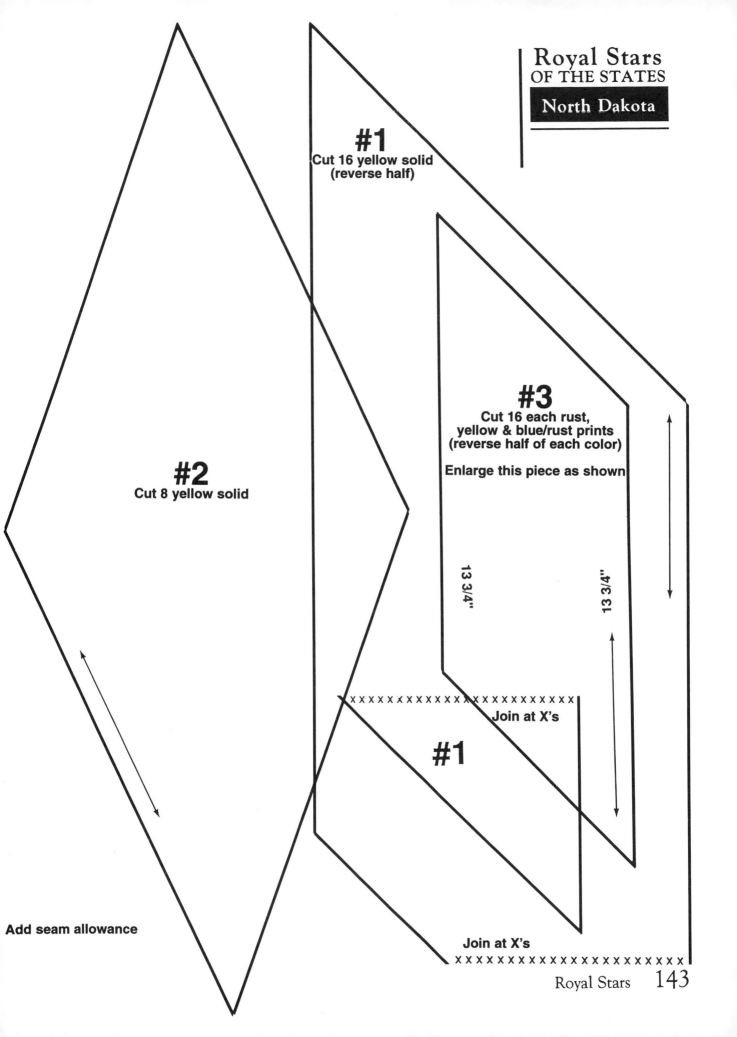

Royal Stars
OF THE STATES
North Dakota

#1
Cut 16 yellow solid
(reverse half)

#2
Cut 8 yellow solid

#3
Cut 16 each rust,
yellow & blue/rust prints
(reverse half of each color)

Enlarge this piece as shown

13 3/4"

13 3/4"

xxxxxxxxxxxxxxxxxxxxxxxx
Join at X's

#1

Add seam allowance

Join at X's
xxxxxxxxxxxxxxxxxxxxxxxxxxx

75" x 75"

MATERIALS
- 1/8 yard wine solid
- 1/4 yard maroon solid
- 1/2 yard maroon print
- 1/2 yard white print
- 3/4 yard blue/white print
- 3/4 yard dark blue print
- 4 yards white solid

PIECING INSTRUCTIONS

1. Cut pieces as directed using measurements given to enlarge pieces #1–#6. Add seam allowance to all pieces before cutting.

2. Sew #2A to #2 to #3 to #4 to #5 to #6 to complete one side of star point. Repeat for reverse pieces.

3. Sew one unit to each side of piece #1 to complete one star point. Repeat for eight points.

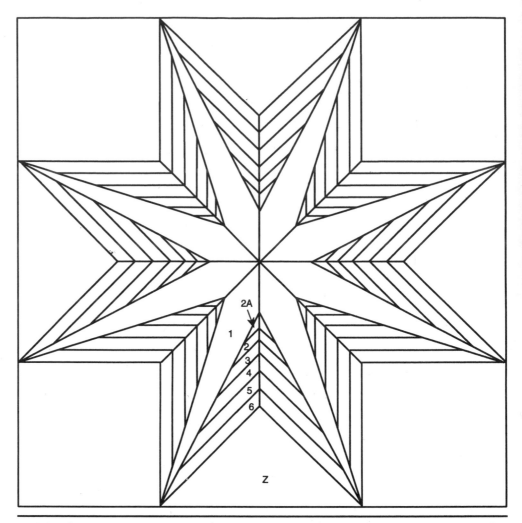

Royal Star of OHIO

4. Join the eight star-point units to make a complete star.

5. Cut four 22" squares for corner fill-in pieces (add seams) from white solid.

6. Cut four Z fill-in triangles (add seams).

7. Set in squares and Z triangles to complete quilt top.

8. Add borders if desired.

See OHIO photo on page 41

ADD
COMMON PIECE Z
(REFER TO PAGES 59–61)

Cut 4 white solid

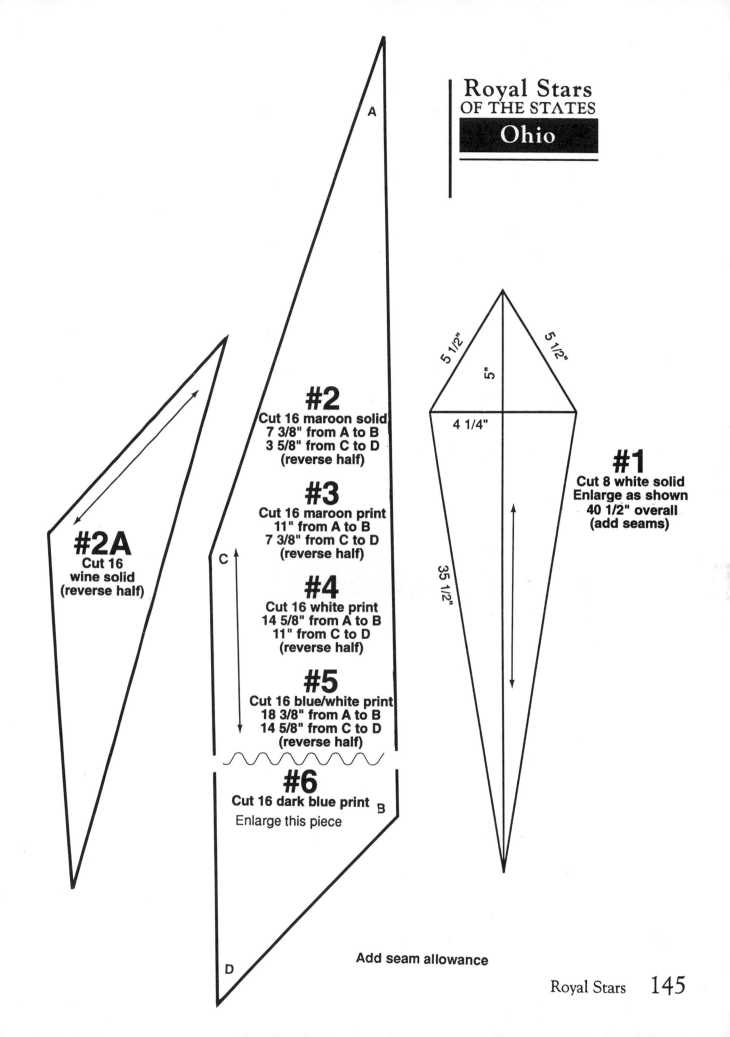

A

5 1/2"

5 1/2"

5"

4 1/4"

#2
Cut 16 maroon solid
7 3/8" from A to B
3 5/8" from C to D
(reverse half)

#3
Cut 16 maroon print
11" from A to B
7 3/8" from C to D
(reverse half)

#1
Cut 8 white solid
Enlarge as shown
40 1/2" overall
(add seams)

C

#4
Cut 16 white print
14 5/8" from A to B
11" from C to D
(reverse half)

35 1/2"

#2A
Cut 16
wine solid
(reverse half)

#5
Cut 16 blue/white print
18 3/8" from A to B
14 5/8" from C to D
(reverse half)

#6
Cut 16 dark blue print

Enlarge this piece

B

D

Add seam allowance

77" x 77"

MATERIALS

- 3/4 yard dark blue print
- 3/4 yard light blue print
- 1 3/4 yards white print
- 1 1/2 yards blue solid
- 3/4 yard white/blue print
- 3 yards white solid

PIECING INSTRUCTIONS

1. Start in the center and piece eight star points using nine B pieces in each.

2. Join the star points to make center star. Set in #2 pieces.

3. Piece the next star point referring to Figure #1. Attach to the center octagon shape.

4. Piece B and C units to attach to #1 referring to Figure #2. Repeat for eight units. Sew to piece #1.

5. Sew the pieced #1-B-C units between the star points to make a large octagon.

6. Sew on the #3 pieces to complete the quilt top.

7. Add borders if desired.

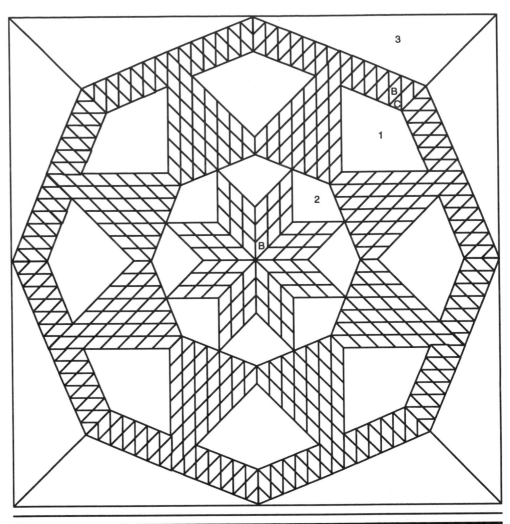

Royal Star of OKLAHOMA

42   **See OKLAHOMA photo on page 42**

ADD
COMMON PIECES B & C
(REFER TO PAGES 59–61)

B   **Cut 64 each dark blue, light blue & white/blue prints, 24 white print & 176 blue solid**

C   **Cut 240 white print**

Add seam allowance

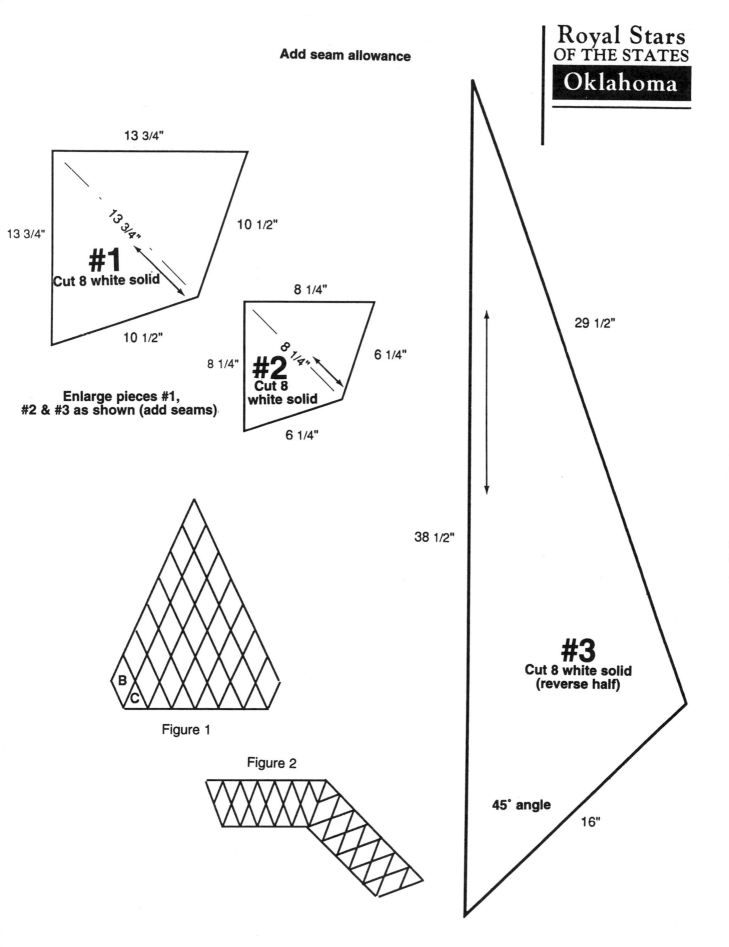

13 3/4"

13 3/4"

13 3/4"

#1
Cut 8 white solid

10 1/2"

10 1/2"

8 1/4"

8 1/4"

8 1/4"

#2
Cut 8
white solid

6 1/4"

6 1/4"

**Enlarge pieces #1,
#2 & #3 as shown (add seams)**

29 1/2"

38 1/2"

B
C

Figure 1

Figure 2

#3
Cut 8 white solid
(reverse half)

45° angle

16"

75" x 75"

MATERIALS
- 3/4 yard gray solid
- 1 yard brown solid
- 3/4 yard orange solid
- 1 1/4 yards rust solid
- 1/2 yard blue solid #1
- 1 yard blue solid #2
- 1/2 yard blue solid #3
- 1/4 yard blue solid #4
- 3 yards white solid

PIECING INSTRUCTIONS

1. Piece one star point beginning in the center with B pieces. Sew nine B pieces together.

2. Sew #6 to #5 to #1. Repeat for reverse pieces.

3. Piece another star point with 25 B pieces. Set the #1-#5-#6 units on each side. Sew #2 to each side of the top point.

4. Set this unit onto the previously pieced B star point. Repeat for eight star points.

5. Join the star points together to complete center.

6. Set in pieces #3 and #4 to complete quilt top.

7. Add borders if desired.

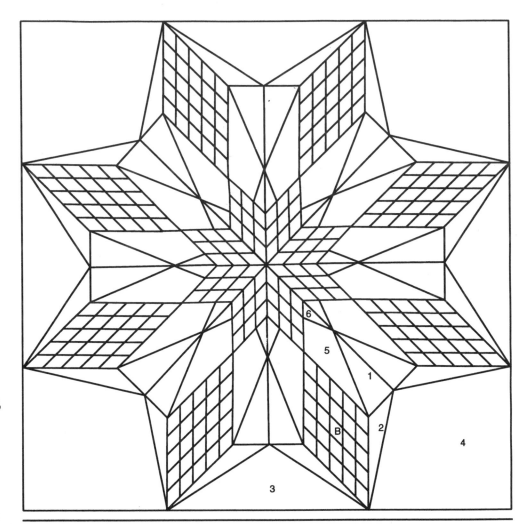

Royal Star of OREGON

See OREGON photo on page 43

ADD
COMMON PIECE B
(REFER TO PAGES 59–61)

B Cut 16 orange, 32 brown, 48 blue #1, 96 blue #2, 56 blue #3 & 24 blue #4

Figure #1

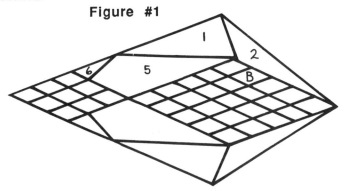

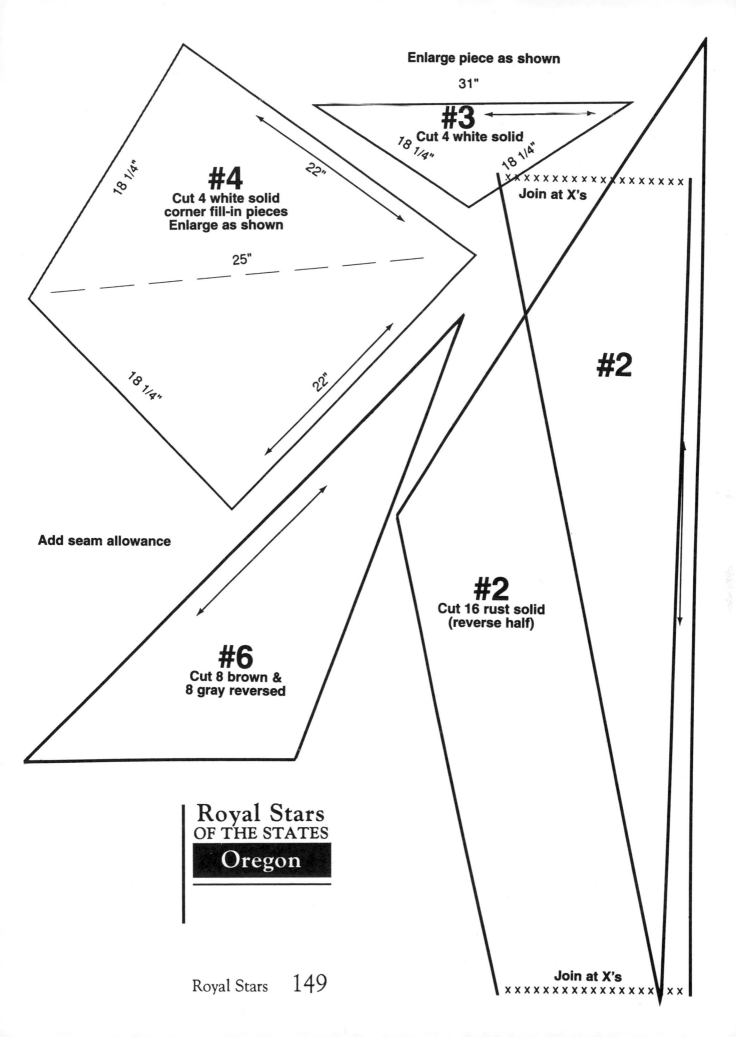

Enlarge piece as shown

31"

#3
Cut 4 white solid

18 1/4"

18 1/4"

Join at X's

18 1/4"

#4
Cut 4 white solid
corner fill-in pieces
Enlarge as shown

22"

25"

18 1/4"

22"

Add seam allowance

#2

#2
Cut 16 rust solid
(reverse half)

#6
Cut 8 brown &
8 gray reversed

Royal Stars
OF THE STATES
Oregon

Join at X's
x x x x x x x x x x x x x x x x x x x x x x

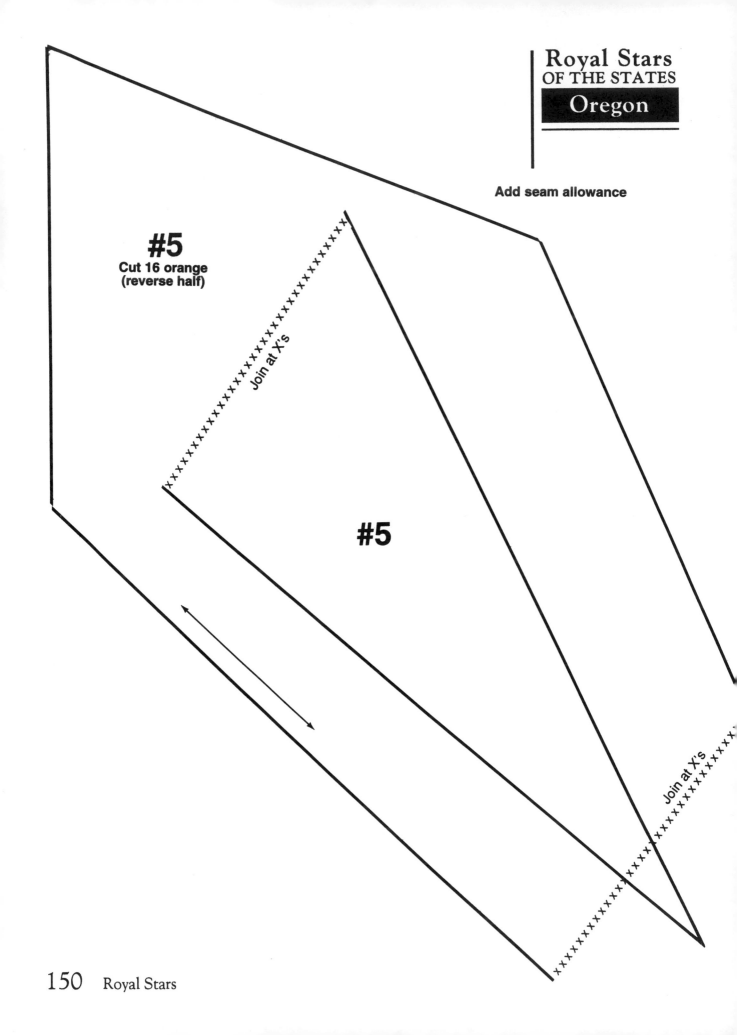

Royal Stars
OF THE STATES
Oregon

Add seam allowance

#5
Cut 16 orange
(reverse half)

Join at X's

#5

Join at X's

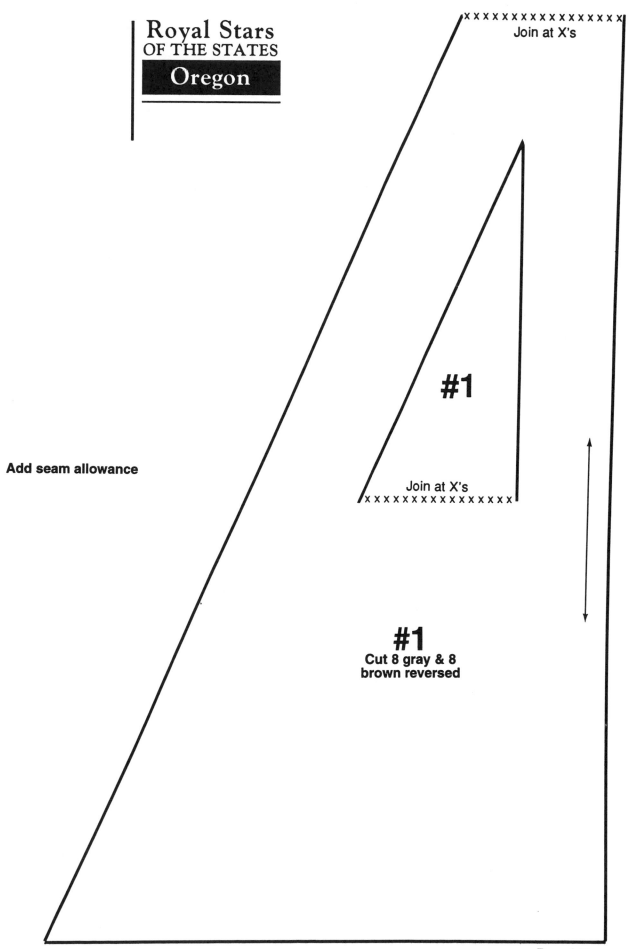

Royal Stars
OF THE STATES
Oregon

Join at X's

#1

Join at X's

Add seam allowance

#1
**Cut 8 gray & 8
brown reversed**

75" x 75"

MATERIALS
- 1/4 yard yellow solid
- 1/4 yard brown print
- 1 1/4 yards green solid
- 1 1/4 yards white print
- 1/4 yard brown solid
- 1 1/4 yards lavender print
- 3 yards white solid

PIECING INSTRUCTIONS

1. To piece one star point, join eight #2 pieces. Set in pieces #6, #7 and #4. Add piece #8 to the #4 sides.

2. Sew three B pieces together and set in C referring to the piecing diagram; repeat. Sew one of these sections to #8.

3. Sew B to #1 to B; repeat. Sew a #1 piece to opposite sides of the previously pieced unit. Sew the B-#1 units to the remaining sides to complete one large star point; repeat for eight points.

4. Join the star points to complete the center star design.

5. Cut four 22" corner squares (add seams) from white solid.

6. Cut four Z fill-in triangles (add seams).

7. Set in the squares and Z triangles to complete the quilt top.

8. Add borders if desired.

Royal Star of PENNSYLVANIA

See PENNSYLVANIA photo on page 44

ADD
COMMON PIECES B, C & Z
(REFER TO PAGES 59–61)

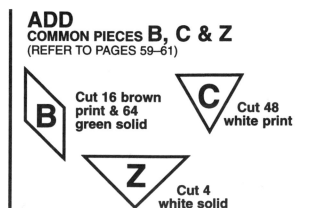

B Cut 16 brown print & 64 green solid

C Cut 48 white print

Z Cut 4 white solid

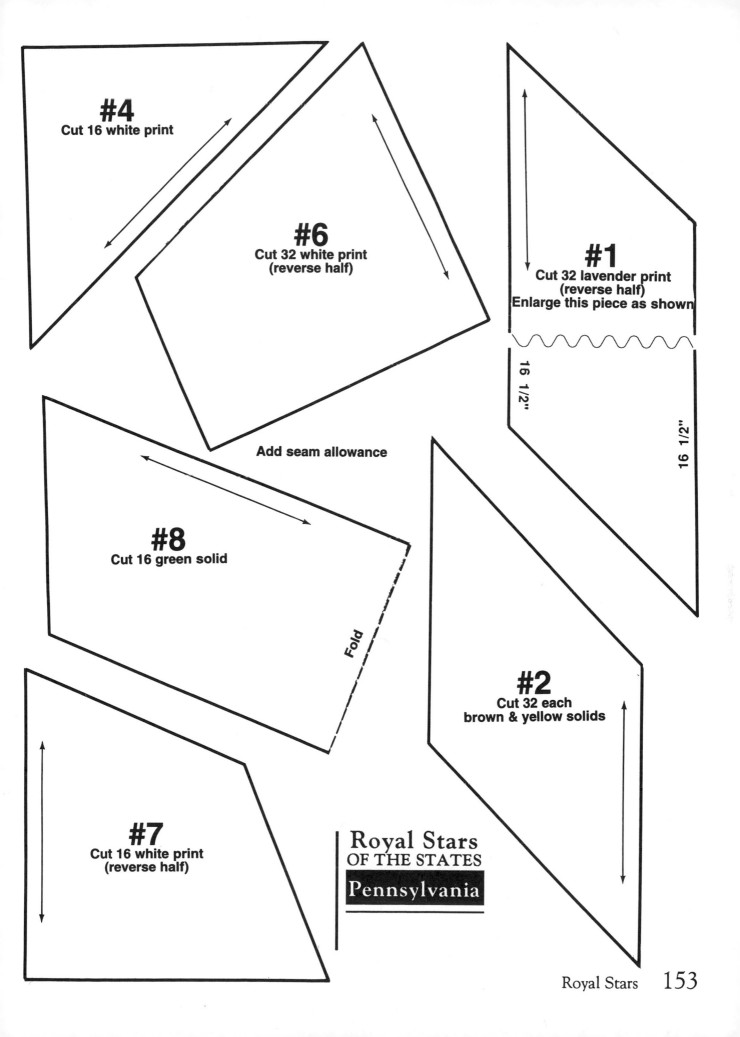

#4
Cut 16 white print

#6
Cut 32 white print
(reverse half)

#1
Cut 32 lavender print
(reverse half)
Enlarge this piece as shown

16 1/2"

16 1/2"

Add seam allowance

#8
Cut 16 green solid

Fold

#2
Cut 32 each
brown & yellow solids

#7
Cut 16 white print
(reverse half)

Royal Stars
OF THE STATES
Pennsylvania

75" x 75"

MATERIALS
- 2 yards red print
- 1/4 yard gray solid
- 1 1/3 yards white-on-white print
- 1 yard red/white print
- 3 yards white solid

PIECING INSTRUCTIONS

1. Piece the small center star using four #3 pieces referring to the piecing diagram; repeat for eight star units.

2. Join the star units to make center star design.

3. Piece a nine-patch square using piece #1; repeat for eight nine-patch units. Set in between center star points.

4. Set #2 pieces between nine-patch units.

5. Sew eight units as shown in Figure #1.

Piece eight units as shown in Figure #2.

6. Sew the pieced units onto the #2 pieces to complete star design.

7. Cut four 22" corner squares (add seams) from white solid.

8. Cut four Z fill-in triangles (add seams).

9. Set in the squares and Z to complete the quilt top.

10. Add borders if desired.

Royal Star of RHODE ISLAND

See RHODE ISLAND photo on page 45

ADD
COMMON PIECES B & Z
(REFER TO PAGES 59–61)

B Cut 40 white-on-white print & 32 red print

Cut 4 white solid

Z

Figure 1

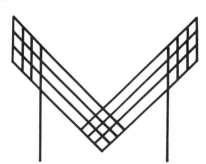

Figure 2

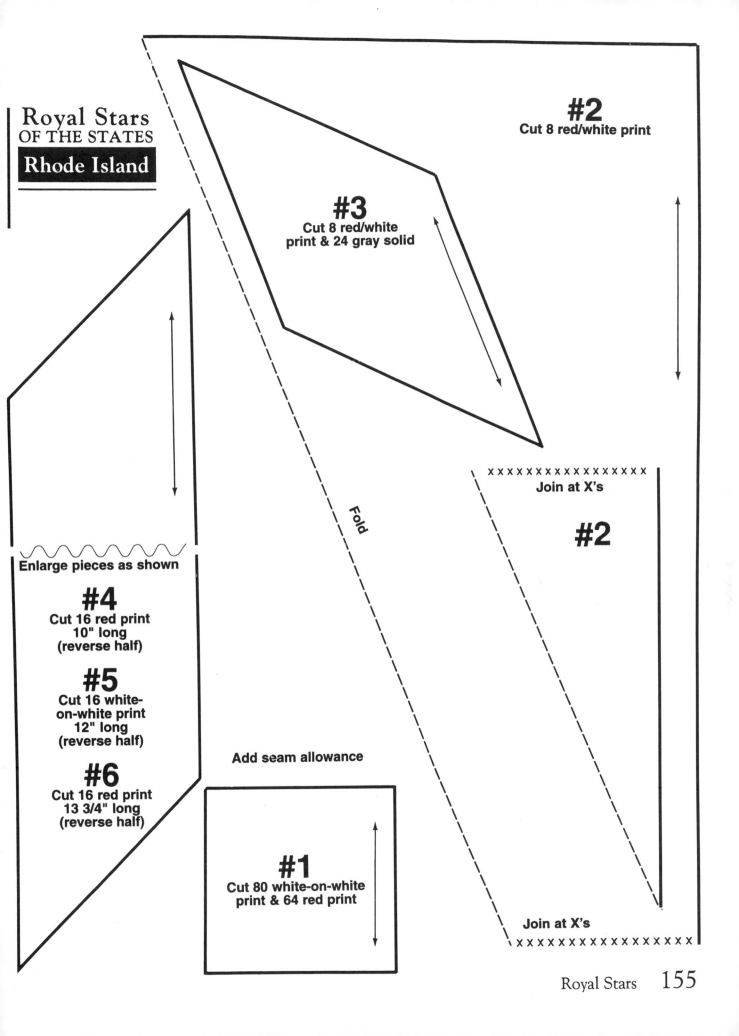

Royal Stars
OF THE STATES
Rhode Island

#2
Cut 8 red/white print

#3
Cut 8 red/white
print & 24 gray solid

x x x x x x x x x x x x x x x x x
Join at X's

#2

Fold

~~~~~~~~~~~~~~~
Enlarge pieces as shown

#4
Cut 16 red print
10" long
(reverse half)

#5
Cut 16 white-
on-white print
12" long
(reverse half)

#6
Cut 16 red print
13 3/4" long
(reverse half)

Add seam allowance

#1
Cut 80 white-on-white
print & 64 red print

Join at X's
x x x x x x x x x x x x x x x x x

72" x 72"

MATERIALS
- 3/4 yard maroon solid
- 1 3/4 yards purple solid
- 3/4 yard navy print
- 1/4 yard white print
- 1/4 yard maroon print
- 3 yards white solid

PIECING INSTRUCTIONS

1. Piece this quilt as a circle starting in the center and working out. Sew all #7 pieces together in a circle.

2. Set in #6 pieces then #5 and continue until all pieces are added for the star design.

3. Set in center piece #8 and appliqué H on top.

4. Cut the A corner pieces and set onto the circle to complete the design.

5. Add borders if desired.

Royal Star of SOUTH CAROLINA

See SOUTH CAROLINA photo on page 46

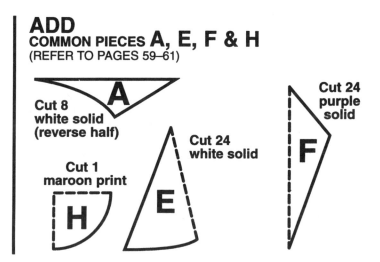

ADD
COMMON PIECES A, E, F & H
(REFER TO PAGES 59–61)

A
Cut 8 white solid (reverse half)

Cut 1 maroon print **H**

Cut 24 white solid **E**

Cut 24 purple solid **F**

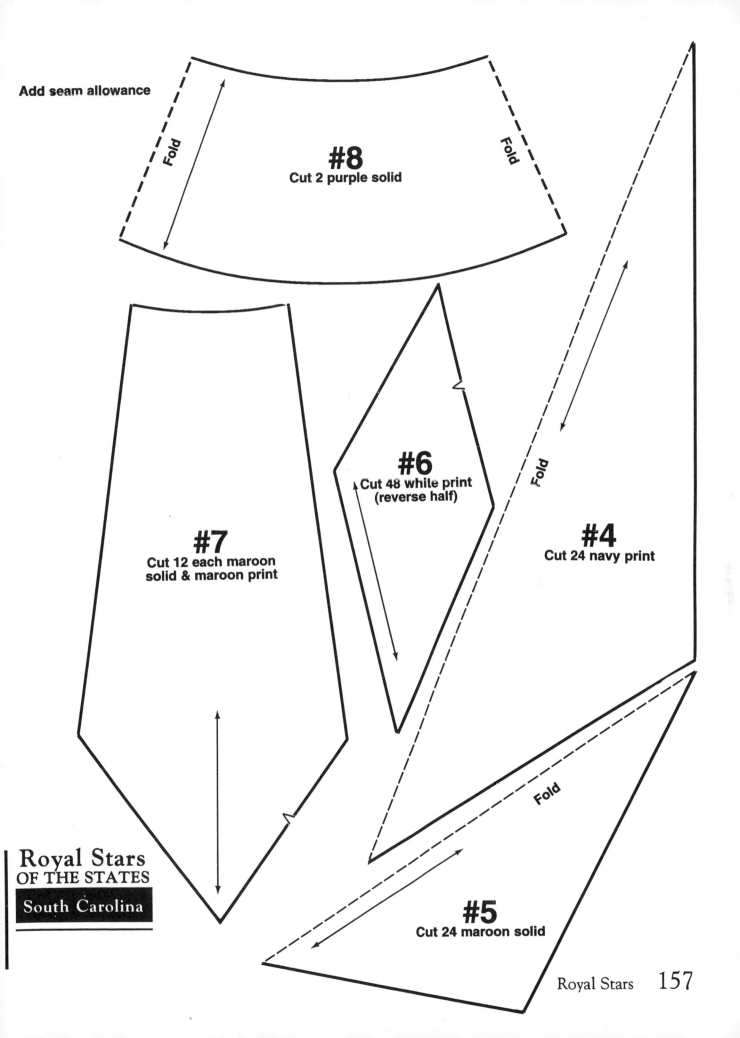

Add seam allowance

Fold

Fold

#8
Cut 2 purple solid

#7
Cut 12 each maroon
solid & maroon print

#6
Cut 48 while print
(reverse half)

Fold

#4
Cut 24 navy print

Fold

#5
Cut 24 maroon solid

Royal Stars
OF THE STATES
South Carolina

75" x 75"

MATERIALS
- 1/8 yard wine solid
- 1/2 yard wine print
- 1 3/4 yards royal blue print
- 1/3 yard white/blue print
- 1/4 yard white print
- 3 yards white solid

PIECING INSTRUCTIONS

Note: This is one of the easiest stars to complete and a good one for beginners to try.

1. Piece eight star points starting in the center with four B diamonds.

2. Add #1 to one side; sew a B diamond to the remaining #1 piece and add to other side. Repeat for pieces #2 through #6. Repeat for eight star points.

3. Sew the star points together to complete the star design.

4. Cut four corner squares, each 22" (add seams), from white solid.

5. Cut four Z fill-in triangles (add seams).

6. Set in the squares and Z triangles to square up quilt.

7. The border pieces given complete a star block as shown in the Border Diagram. The yardage requirements given are for 12 blocks (six each on the top and bottom). If blocks will be added to all sides, increase yardage as necessary.

8. To piece one border block, sew all #1 pieces together to make star center; set in pieces #2 and #3 to complete a block. Repeat for desired numbers of blocks.

9. Add borders to quilt top and bottom to complete.

Royal Star of SOUTH DAKOTA

See SOUTH DAKOTA photo on page 47

ADD
COMMON PIECES B & Z
(REFER TO PAGES 59–61)

B

Cut 56 wine print, 8 wine solid & 16 royal blue print

Cut 4 white solid

Z

#2
Cut 48 for borders

Border Yardage
Needed for Each Piece:
1/2 yard each 2 colors for #1; 1/2
yard for #2; and 1/2 yard for #3

Border Diagram

#1
Cut 16 white print
5 1/2" long
(reverse half)

#2
Cut 16 royal blue print
8 1/4" long
(reverse half)

#3
Cut 16 white/blue print
11" long
(reverse half)

#4
Cut 16 royal blue solid
13 3/4" long
(reverse half)

#5
Cut 16 white/blue print
16 1/2" long
(reverse half)

#6
Cut 16 royal blue print
19 1/4" long
(reverse half)

Enlarge this piece

#3
Cut 48 for borders

#1
Cut 48 each of 2
colors for borders

Royal Stars
OF THE STATES
South Dakota

Add seam allowance

72" x 72"

MATERIALS

- 1 yard black
- 1/2 yard rust
- 1/4 yard hot pink
- 1/3 yard wine
- 1/3 yard fuchsia
- 1/2 yard each blue #3, #4 and #5
- 1/3 yard purple
- 2/3 yard each blue #1 and #2
- 3/4 yard dark green
- 3 yards white solid

PIECING INSTRUCTIONS

1. Piece one bar section using pieces #2 through #11; repeat for reverse pieces.

2. Join the two pieced sections down the center seam to create one star section; repeat for 12 sections.

3. Sew a #13 piece to each side of #12; repeat for 12 sections.

4. Join the pieced bar points with the #12-#13 units to complete the star center.

5. Sew the A pieces around the outside of the circle shape to complete star top.

6. Borders may be added as desired.

Royal Star of TENNESSEE

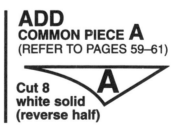

See TENNESSEE photo on page 48

ADD
COMMON PIECE A
(REFER TO PAGES 59–61)

A

Cut 8 white solid (reverse half)

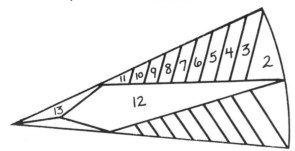

Figure 1

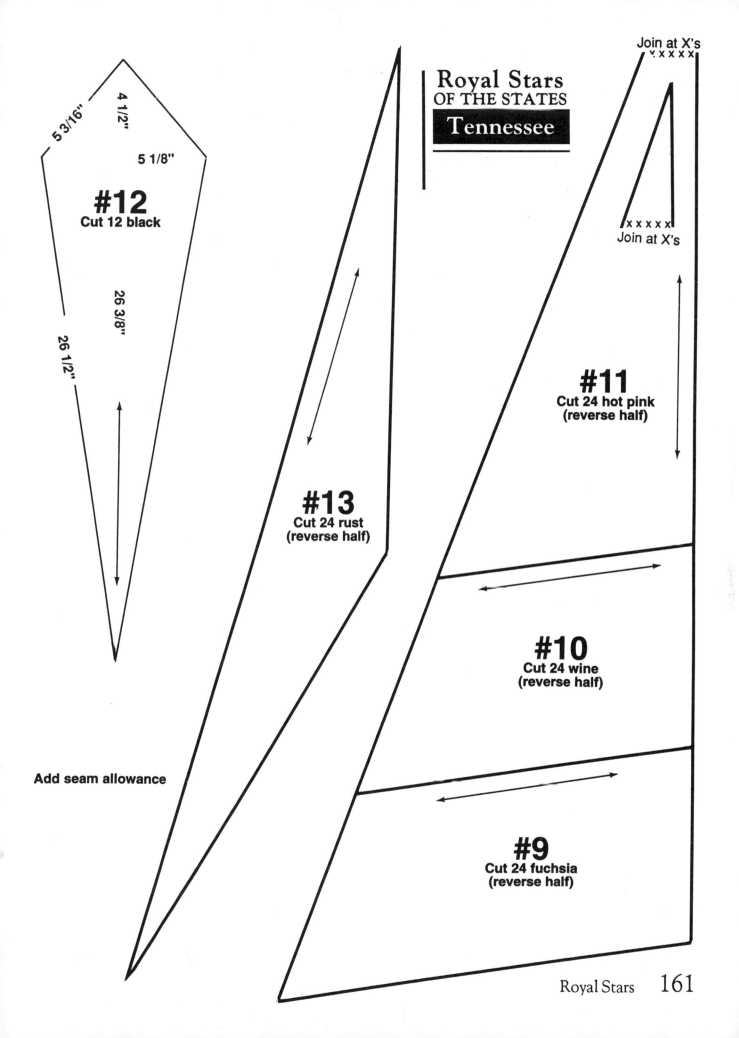

Join at X's
x x x x x

Royal Stars
OF THE STATES
Tennessee

x x x x x
Join at X's

4 1/2"

5 3/16"

5 1/8"

#12
Cut 12 black

26 3/8"

26 1/2"

#11
Cut 24 hot pink
(reverse half)

#13
Cut 24 rust
(reverse half)

#10
Cut 24 wine
(reverse half)

Add seam allowance

#9
Cut 24 fuchsia
(reverse half)

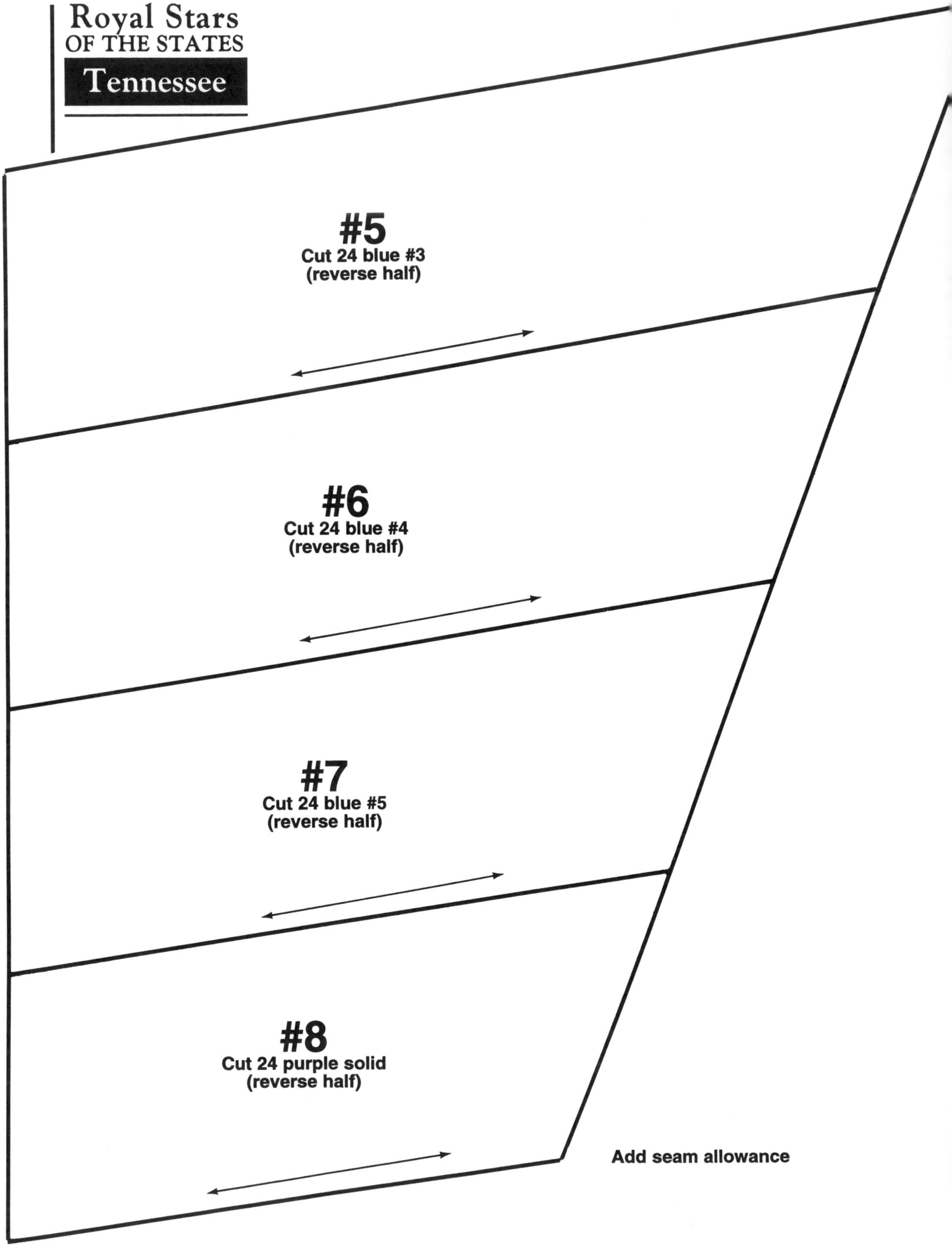

#5
Cut 24 blue #3
(reverse half)

#6
Cut 24 blue #4
(reverse half)

#7
Cut 24 blue #5
(reverse half)

#8
Cut 24 purple solid
(reverse half)

Add seam allowance

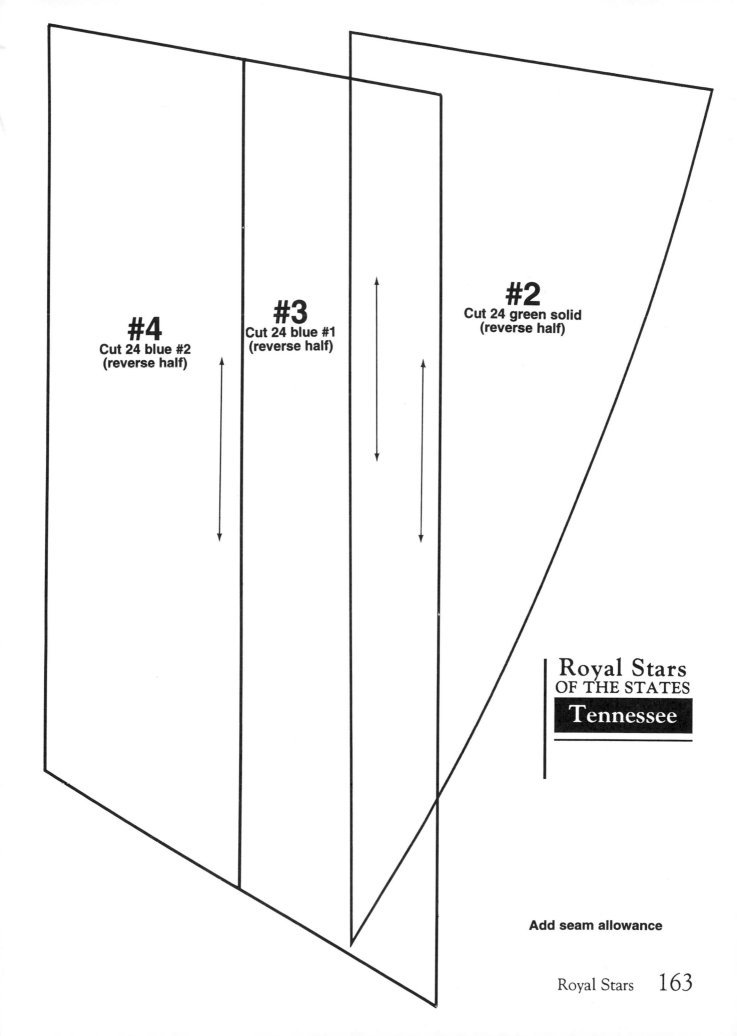

#4
Cut 24 blue #2
(reverse half)

#3
Cut 24 blue #1
(reverse half)

#2
Cut 24 green solid
(reverse half)

Royal Stars
OF THE STATES
Tennessee

Add seam allowance

75" x 75"

MATERIALS
- 3/4 yard red solid
- 1 yard light blue print
- 2/3 yard blue print
- 1 1/2 yards navy solid
- 3 yards white solid

PIECING INSTRUCTIONS

1. Piece one star section beginning in the center. Join seven B pieces with 12 C pieces. Sew a #4 piece to each C side.

2. Sew four B's and eight C's; sew #3 to one C side and #2 to the other C side; repeat. Sew two C's to B; sew #1 to the side; repeat.

3. Join the pieced units referring to the drawing of one star point. Sew a #5 piece to each side and B at the point to finish; repeat for eight star sections.

4. Join the star sections to complete the center.

5. Cut four 22" corner squares (add seams) from white solid.

6. Cut four Z fill-in triangles.

7. Set in the Z triangles and corner squares to finish quilt top.

8. Borders may be added if desired.

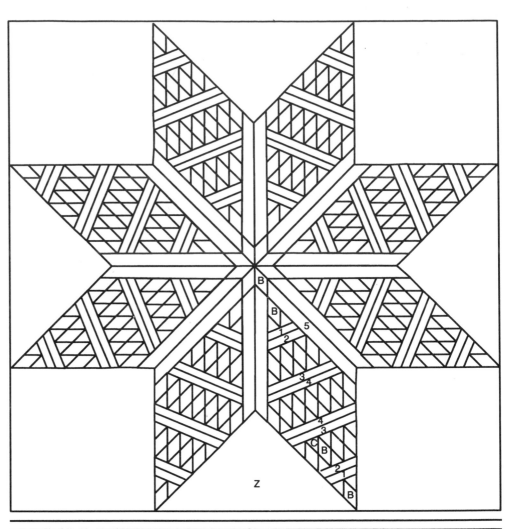

Royal Star of TEXAS

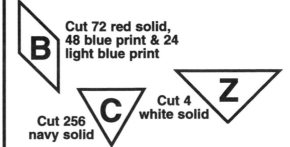

See TEXAS photo on page 49

ADD
COMMON PIECES B, C & Z
(REFER TO PAGES 59–61)

B Cut 72 red solid, 48 blue print & 24 light blue print

C Cut 256 navy solid

Z Cut 4 white solid

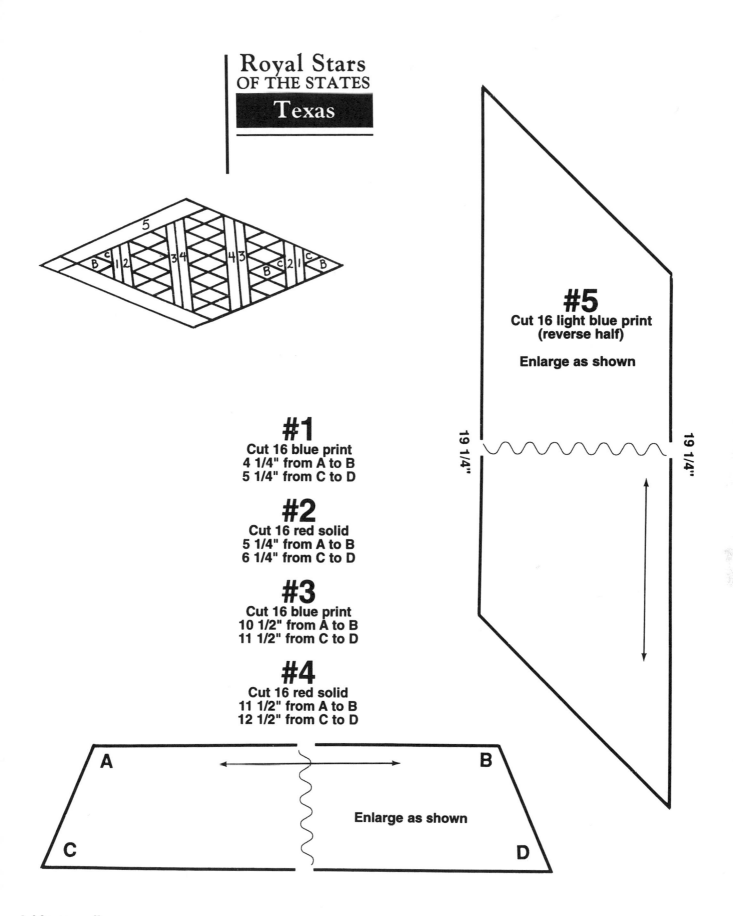

Royal Stars
OF THE STATES
Texas

#5
Cut 16 light blue print
(reverse half)

Enlarge as shown

19 1/4"

19 1/4"

#1
Cut 16 blue print
4 1/4" from A to B
5 1/4" from C to D

#2
Cut 16 red solid
5 1/4" from A to B
6 1/4" from C to D

#3
Cut 16 blue print
10 1/2" from A to B
11 1/2" from C to D

#4
Cut 16 red solid
11 1/2" from A to B
12 1/2" from C to D

A

B

Enlarge as shown

C

D

Add seam allowance

75" x 75"

MATERIALS
- 1/2 yard light blue
- 3/4 yard dark blue
- 1/2 yard light brown
- 1/2 yard light rust
- 2/3 yard mustard
- 3/4 yard dark rust
- 1 yard cream
- 3 yards white solid

PIECING INSTRUCTIONS
1. Piece bar sections with pieces #2 through #7, sewing half of the star section at a time. Start with pieces marked with X's in Figure #1.

2. Join the two pieced sections and set in piece #1 as shown in Figure #1. Repeat for eight star sections.

3. Join the star sections to complete star design.

4. Cut four 22" corner squares (add seams) from white solid.

5. Cut four Z fill-in triangles.

6. Set in the Z triangles and corner squares to finish quilt top.

7. Borders may be added if desired.

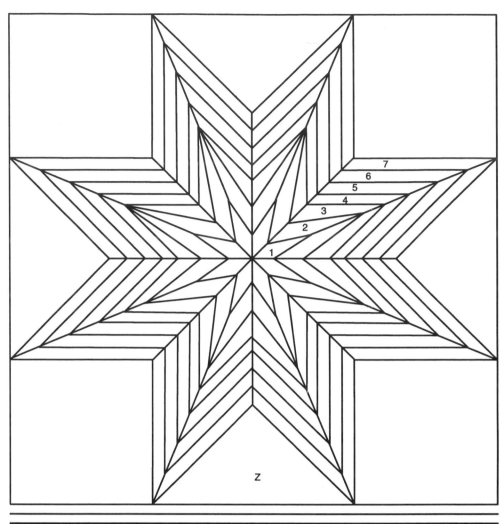

Royal Star of UTAH

50

See UTAH photo on page 50

ADD
COMMON PIECE Z
(REFER TO PAGES 59–61)

Cut 4 white solid

Z

Join at X's
× × × × ×

Join at X's
× × × × ×

#1
Cut 8 light blue

Add seam allowance

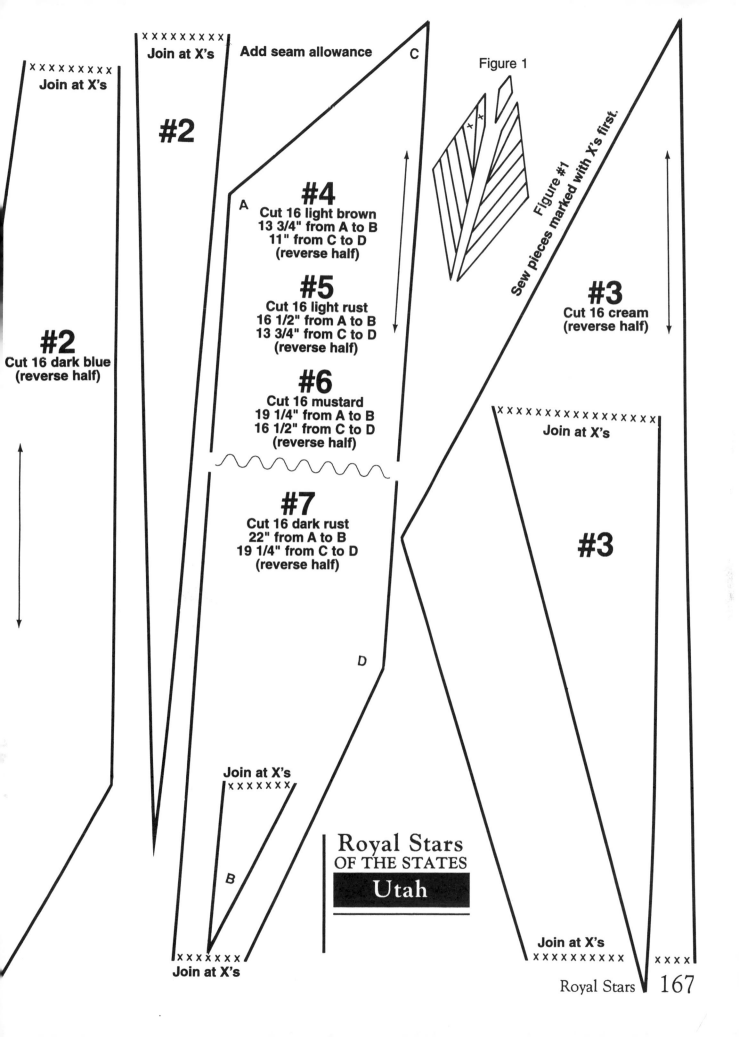

#2

#2
Cut 16 dark blue
(reverse half)

A

#4
Cut 16 light brown
13 3/4" from A to B
11" from C to D
(reverse half)

#5
Cut 16 light rust
16 1/2" from A to B
13 3/4" from C to D
(reverse half)

#6
Cut 16 mustard
19 1/4" from A to B
16 1/2" from C to D
(reverse half)

#7
Cut 16 dark rust
22" from A to B
19 1/4" from C to D
(reverse half)

Figure #1
Sew pieces marked with X's first.

#3
Cut 16 cream
(reverse half)

Join at X's

#3

Join at X's
xxxxxxx

Royal Stars
OF THE STATES
Utah

B

Join at X's

Join at X's
xxxxxxxxxxx

75" x 75"

MATERIALS
- 1/2 yard each aqua, purple, forest green and green
- 3/4 yard royal blue
- 1/4 yard each pink, wine, coral and maroon
- 1 yard black
- 3 yards white solid

PIECING INSTRUCTIONS

1. Piece one star segment beginning in the center with four B diamonds. Set on #1.

2. Set in #2.

3. Sew fifteen B diamonds to make a section referring to Figure #1. Sew to one side of piece #2.

4. Sew another B section with 24 diamonds, referring to Figure #1. Sew to remaining side of piece #2 to complete the star section.

5. Repeat for eight sections; join the sections to complete the star center.

6. Cut four 22" corner squares (add seams) from white solid.

7. Cut four Z fill-in triangles.

8. Set in the Z triangles and corner squares to finish quilt top.

9. Borders may be added if desired.

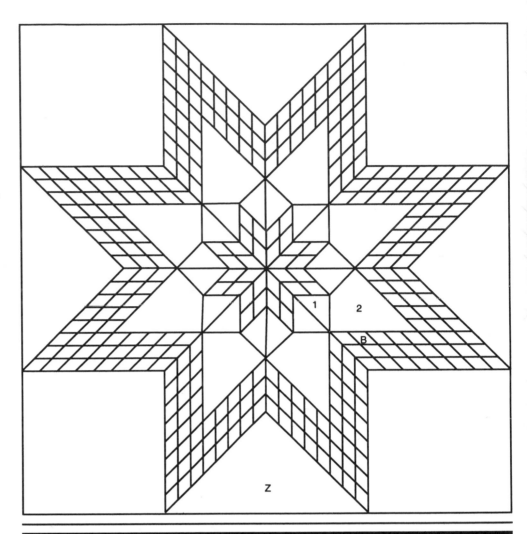

Royal Star of VERMONT

See VERMONT photo on page 51

ADD COMMON PIECES B & Z
(REFER TO PAGES 59–61)

B — Cut 48 aqua, 16 pink, 32 wine, 48 purple, 48 forest green, 48 green, 32 coral, 24 maroon & 48 royal blue

Z — Cut 4 white solid

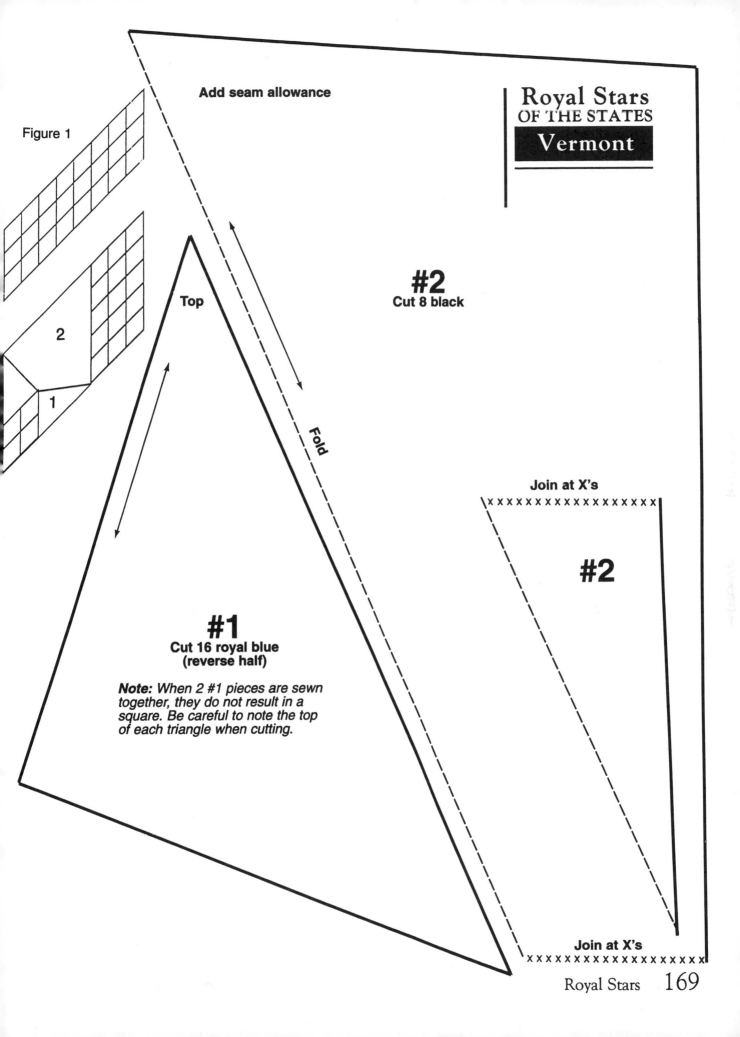

Add seam allowance

Figure 1

2

1

Top

#2
Cut 8 black

Fold

Join at X's
x x x x x x x x x x x x x x x x x x x

#2

#1
Cut 16 royal blue
(reverse half)

Note: *When 2 #1 pieces are sewn together, they do not result in a square. Be careful to note the top of each triangle when cutting.*

Join at X's
x x x x x x x x x x x x x x x x x x x x x x x

72" x 72"

MATERIALS
- 1 3/4 yards rust print
- 3/4 yard rust/blue print
- 1/2 yard dark blue print
- 1/4 yard light blue print
- 1/4 yard beige print
- 1/4 yard yellow solid
- 1/4 yard cream solid
- 3 yards white solid

PIECING INSTRUCTIONS

Note: This design cannot be easily completed in star points or sections.

1. Starting in the center, sew all #8 pieces together in a circle.

2. Set in all #7 pieces. Continue adding pieces around the center in descending numerical order.

3. Set in pieces F and then E to complete pieced circle.

4. Appliqué piece H over center.

5. Sew A pieces to circle to complete the quilt top.

6. Borders may be added if desired.

Royal Star of VIRGINIA

See VIRGINIA photo on page 52

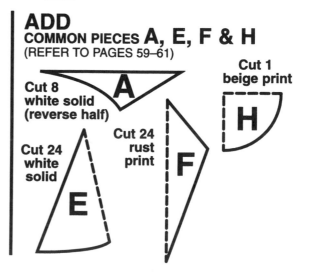

ADD
COMMON PIECES A, E, F & H
(REFER TO PAGES 59–61)

Cut 8 white solid (reverse half) — **A**

Cut 24 white solid — **E**

Cut 24 rust print — **F**

Cut 1 beige print — **H**

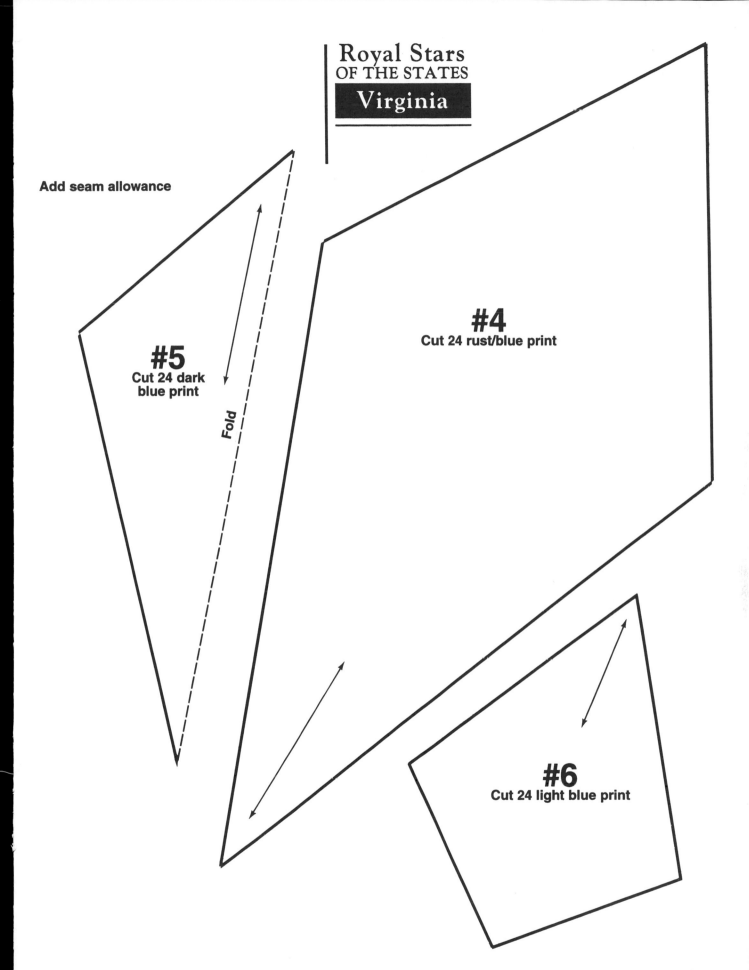

Royal Stars
OF THE STATES
Virginia

Add seam allowance

#5
Cut 24 dark
blue print

Fold

#4
Cut 24 rust/blue print

#6
Cut 24 light blue print

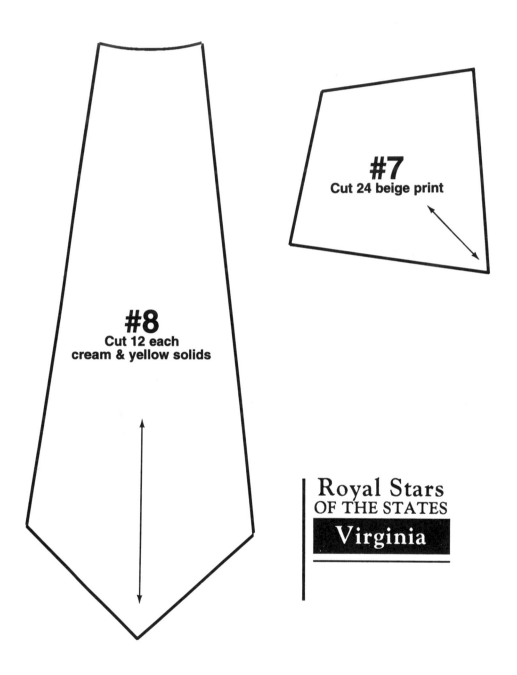

#7
Cut 24 beige print

#8
Cut 12 each
cream & yellow solids

Royal Stars
OF THE STATES
Virginia

72" x 72"

MATERIALS
- 1/2 yard cream solid
- 1/8 yard blue #1
- 1/3 yard blue #2
- 1/2 yard each blue #3 and #4
- 1 1/4 yards orange solid
- 1/2 yard each pink #1 and #2
- 3/4 yard pink #3
- 3 yards white solid

PIECING INSTRUCTIONS
1. Referring to Figure #1, sew pieces #2, #5, #6 and #7 together and sew to #3. Repeat for reverse pieces; join the two sections.

2. Sew piece #8 to #9 to #10 to #11; repeat for reverse pieces. Join together and set in piece #4.

3. Set the #4 section into the #3 section to complete a star unit as shown in Figure #1. Repeat for 12 units.

4. Join the units to complete the star design.

5. Appliqué piece I over center.

6. Sew A pieces to circle to complete the quilt top.

7. Borders may be added if desired.

Royal Star of WASHINGTON

See WASHINGTON photo on page 53

ADD
COMMON PIECES A & I
(REFER TO PAGES 59–61)

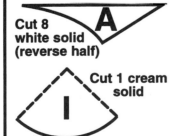

A

Cut 8
white solid
(reverse half)

Cut 1 cream
solid

I

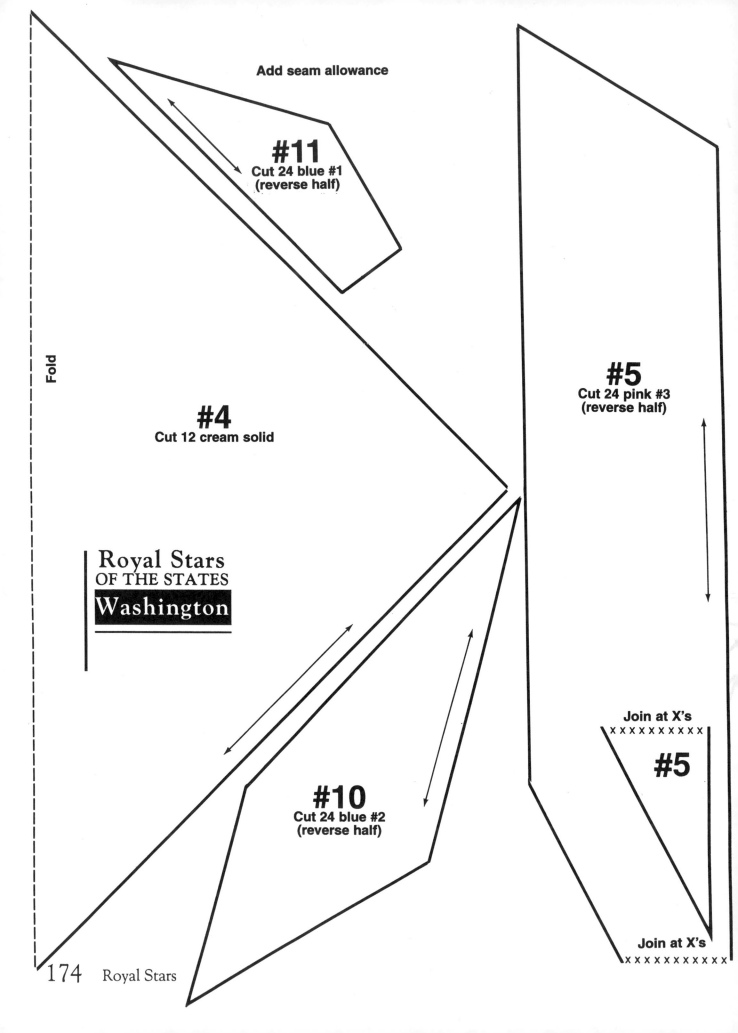

Add seam allowance

#11
Cut 24 blue #1
(reverse half)

Fold

#4
Cut 12 cream solid

Royal Stars
OF THE STATES
Washington

#5
Cut 24 pink #3
(reverse half)

Join at X's
x x x x x x x x x x x

#5

#10
Cut 24 blue #2
(reverse half)

Join at X's
x x x x x x x x x x x

Add seam allowance

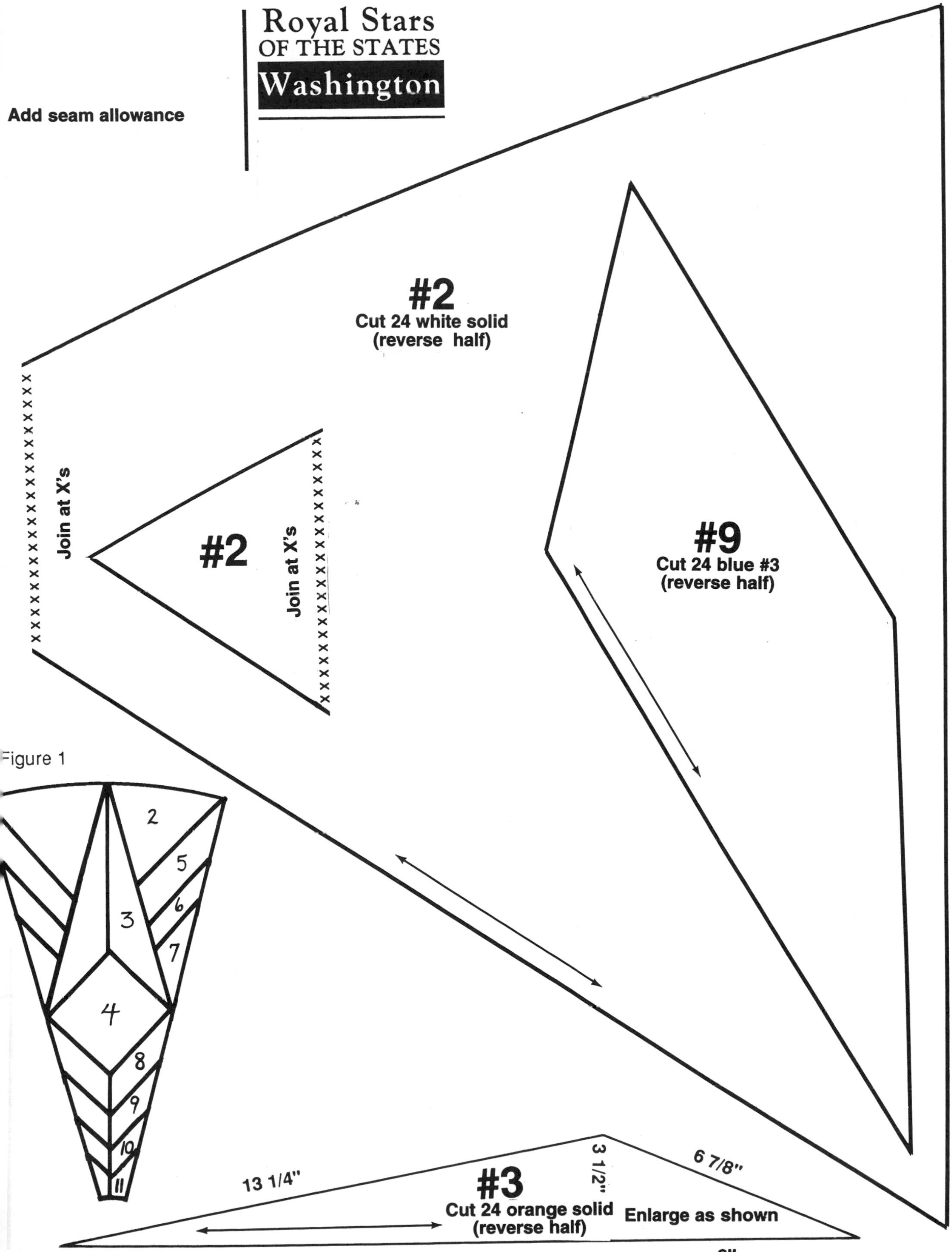

#2
Cut 24 white solid
(reverse half)

Join at X's

#2

Join at X's

#9
Cut 24 blue #3
(reverse half)

Figure 1

2
5
3
6
7
4
8
9
10
11

13 1/4"

3 1/2"

6 7/8"

#3
Cut 24 orange solid
(reverse half)   **Enlarge as shown**

12 3/4"

6"

Royal Stars   175

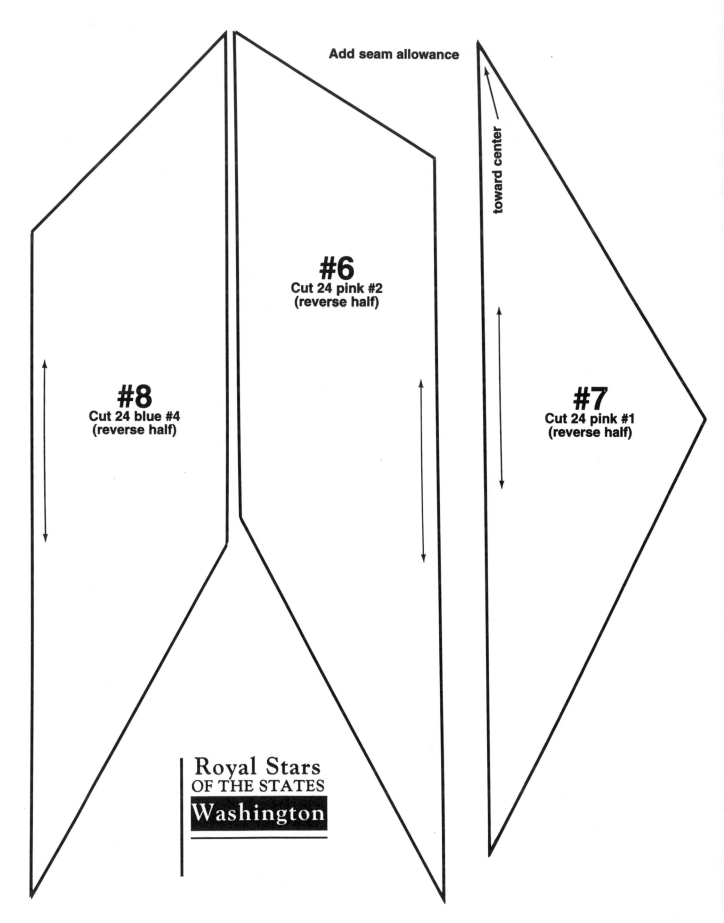

Add seam allowance

toward center

#6
Cut 24 pink #2
(reverse half)

#8
Cut 24 blue #4
(reverse half)

#7
Cut 24 pink #1
(reverse half)

Royal Stars
OF THE STATES
Washington

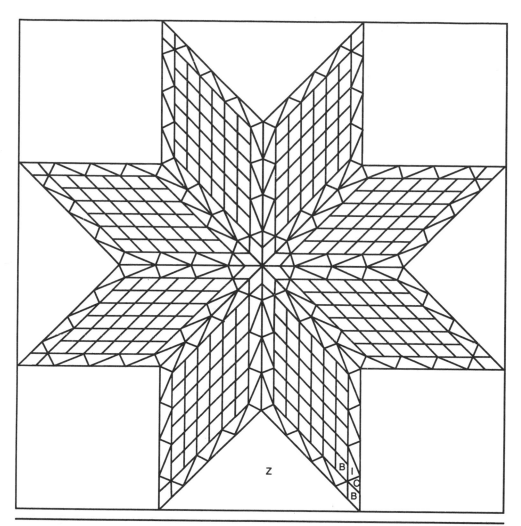

75" x 75"

MATERIALS
- 1/2 yard maroon print
- 3/4 yard brown print
- 3/4 yard cream solid
- 1/8 yard each pale yellow, yellow, blue #3 and blue #4
- 1/4 yard each pale pink, lavender, blue #1 and blue #2
- 1/2 yard each light purple, purple and plum
- 3 yards white solid

PIECING INSTRUCTIONS

1. Piece one star point with 36 B diamonds. Refer to color photo of quilt on page 54 for color placement.

2. Sew six #1 pieces to each side of the B star unit.

3. Sew two C pieces to the base of one B; repeat. Sew one of these units to each end of the pieced unit to complete one star section.

4. Repeat for eight star sections.

5. Join the star sections to complete star design.

6. Cut four 22" squares for corners (add seams).

7. Cut four Z fill-in triangles for sides (add seams), from white solid.

8. Set in squares and Z triangles to complete star design.

9. Borders may be added if desired.

Royal Star of WEST VIRGINIA

See WEST VIRGINIA photo on page 54

ADD
COMMON PIECES B, C & Z
(REFER TO PAGES 59–61)

B Cut 8 pale yellow, 16 yellow, 24 pale pink, 32 lavender, 40 light purple, 48 purple, 40 plum, 32 blue #1, 24 blue #2, 16 blue #3, 8 blue #4 & 16 brown print

C Cut 16 each cream & maroon print

Z Cut 4 white solid

Add seam allowance

#1 Cut 48 maroon print, 96 brown print & 48 cream (reverse half)

75" x 75"

MATERIALS

- 2 1/2 yards light blue print
- 1/2 yard white/blue print
- 1/2 yard blue print
- 1 yard violet solid
- 3/4 yard blue pin dot
- 1 yard navy pin dot
- 1/8 yard white print
- 3 yards white solid

PIECING INSTRUCTIONS

1. Sew center star sections using nine B diamonds; repeat to make eight star sections.

2. Join the star sections to make center star. Set in pieces #1 and #2.

3. Sew #3 strips to each side.

4. Sew #4 strips to #5; repeat for four units. Sew onto pieced section.

Royal Star of WISCONSIN

5. Piece the outside star points with 28 B diamonds. Sew C pieces to the end of each B strip before joining to make star point; repeat for eight units.

6. Set a star unit onto each #4 piece to complete the star design.

7. Cut four 22" squares for corners (add seams) from white solid.

8. Cut four Z fill-in triangles for sides (add seams).

9. Set in squares and Z triangles to complete star design.

10. Eight-pointed star units may be pieced and appliquéd to the corner squares, if desired. Use piece B in colors of your choice to match center.

11. Borders may be added if desired.

55

See WISCONSIN photo on page 55

ADD
COMMON PIECES B, C & Z
(REFER TO PAGES 59–61)

B Cut 32 white/blue print, 56 blue print, 120 violet solid, 64 blue pin dot, 8 white print & 16 light blue print

C Cut 64 navy pin dot

Z Cut 4 white solid

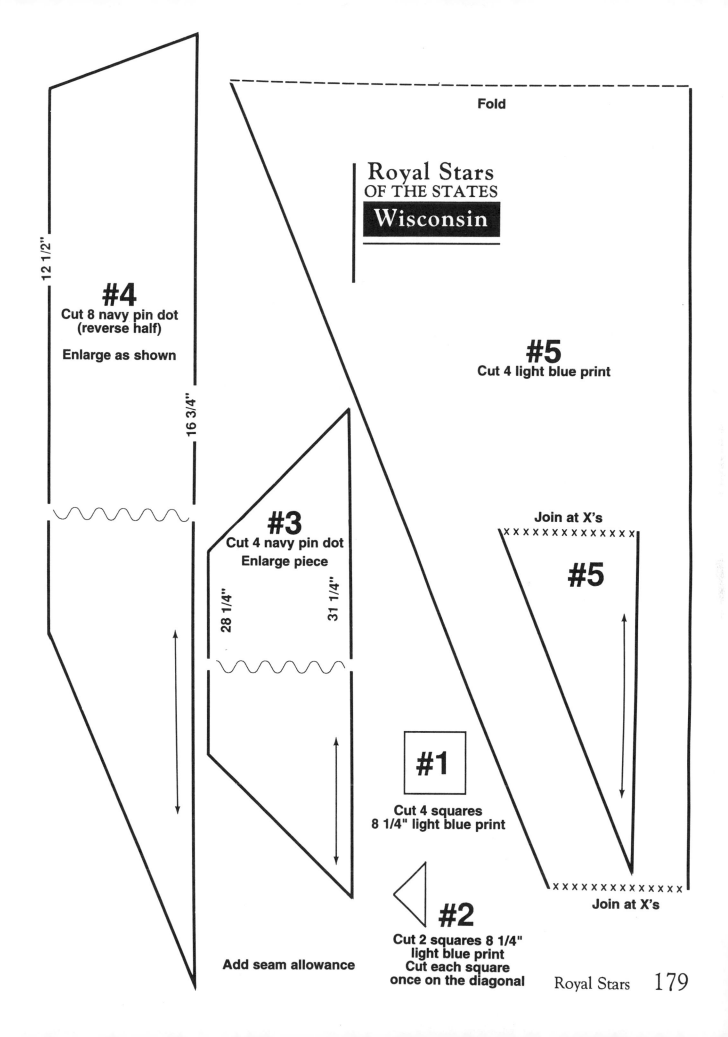

Fold

Royal Stars
OF THE STATES
Wisconsin

12 1/2"

#4
Cut 8 navy pin dot
(reverse half)

Enlarge as shown

16 3/4"

#5
Cut 4 light blue print

Join at X's
x x x x x x x x x x x x x x

#5

#3
Cut 4 navy pin dot
Enlarge piece

28 1/4"

31 1/4"

#1
Cut 4 squares
8 1/4" light blue print

x x x x x x x x x x x x x x x
Join at X's

#2
Cut 2 squares 8 1/4"
light blue print
Cut each square
once on the diagonal

Add seam allowance

75" x 75"

MATERIALS
- 3/4 yard purple solid
- 1/2 yard lavender print
- 1/2 yard light blue print
- 3/4 yard black
- 1 1/4 yards light blue solid
- 1/2 yard dark blue solid
- 1/8 yard dark blue print
- 3 yards white solid

PIECING INSTRUCTIONS

1. Sew two C's to the base of one B 10 times. Sew a C to the bottom right side of one B; repeat with C on the bottom left side of another B. Sew two units with C on the top right and bottom left of B and two units with C on the top left and bottom right of B.

2. Arrange the pieced units with #1 and join in rows to make one star point, referring to color photo on page 56 for color placement.

3. Sew piece #2 on each side of the star point; repeat for eight star units.

4. Join the star units to complete the star design.

5. Cut four 22" squares for corners (add seams) from white solid.

6. Cut four Z fill-in triangles for sides (add seams).

7. Set in squares and Z triangles to complete star design.

8. Borders may be added if desired.

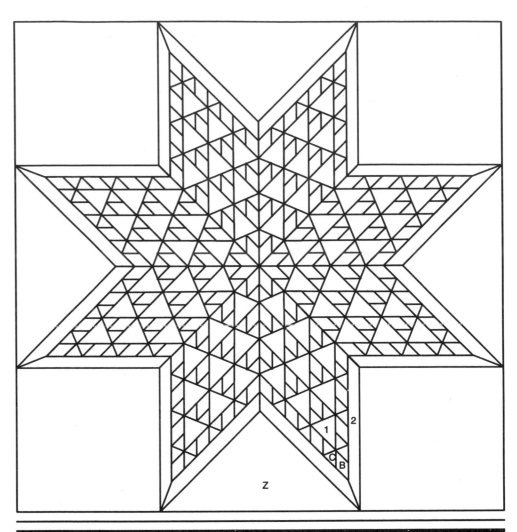

Royal Star of WYOMING

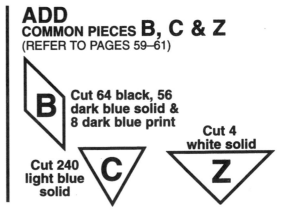

See WYOMING
photo on page 56

ADD
COMMON PIECES B, C & Z
(REFER TO PAGES 59–61)

B Cut 64 black, 56 dark blue solid & 8 dark blue print

Cut 240 light blue solid **C**

Cut 4 white solid **Z**

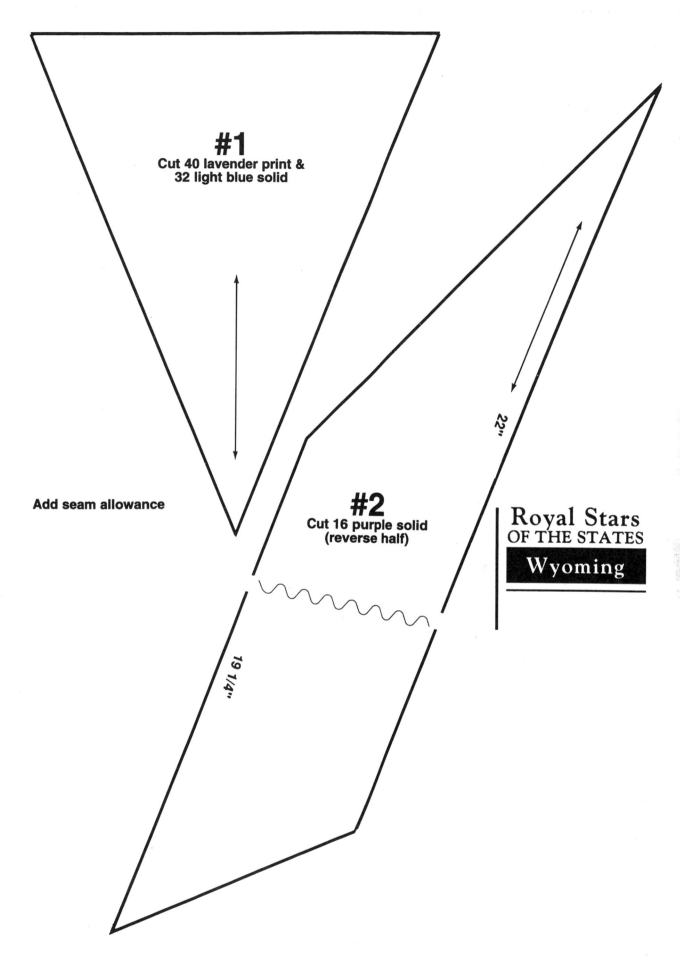

#1
Cut 40 lavender print &
32 light blue solid

Add seam allowance

#2
Cut 16 purple solid
(reverse half)

22"

19 1/4"

Royal Stars
OF THE STATES
Wyoming

orders may be added to any of the Royal Star designs to complement the center design and make a larger quilt. The easiest border treatment includes adding fabric strips of different widths to the outside edges. The strips may be butted (Figure 1) or mitered (Figure 2) at the corners.

If you need a longer quilt, add borders to the top and bottom only. Most of the Royal Star quilt designs finish to either a 72" or 75" square. These sizes may not be perfect for your size bed. Adding borders can make the quilt almost any size you need.

Several of the Royal Stars designs included simple border patterns with a few instructions for using them on the quilt they are included with.

Almost all of Dolores Yoder's quilts shown in the color section of this book include a pieced border. Several of the borders are repeated on more than one quilt.

We have chosen five of Dolores' designs to share with you here. Several are given for both the 72" and 75" size. Because a border design can be used on more than one quilt, color suggestions are not given. Yardage is given for each piece as needed to complete the border on the size quilt specified.

Use colors that match the colors used in your quilt in the border design. After planning the colors, refer to the list of materials given with each border design for the amount of fabric needed for each template (except where corner pieces are included with side pieces using the same fabric).

A drawing showing the border unit with corners and templates with the number of pieces to cut are given with brief instructions for piecing each border design.

Borders given for a 72" quilt will fit any quilt of that size; however, if you would like to use a 75" border design on a 72" quilt, it can be done very easily. Simply add 1 1/2" border strips (cut 2" wide) on each side of the finished quilt to enlarge it to 75".

If you would like to use a 72" border design on a 75" quilt, determine the size of the pieces that run along the quilt edge. For example, Border 1 has a triangle with a 4" diagonal along the inside edge. Eighteen triangles will fit along the edge on a 72" quilt.

If you want to use this border on a 75" quilt, you will need a number that can be divided by 4—76 would work. Add a 1" border all around a 75" quilt and cut 19 border triangles for each side.

Using these methods, most of the border designs can be made to fit any quilt size.

The templates are given without seam allowance. When cutting, add 1/4" all around for a seam allowance. Refer to the general instructions on page 57 for piecing.

Borders add much more to a quilt than just size. They frame the center design and balance the colors. Although any of the Royal Star designs may be made without borders, they look much better when borders are added. Choose a design from among those given here, from another source, or design one of your own and make your Royal Star quilt even more beautiful!

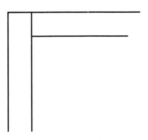

Figure 1
Borders may be butted
at the corners.

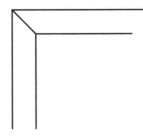

Figure 2
Borders may be mitered at
the corners.

This border design was used on the Royal Stars of Connecticut and Florida, both 72" quilts. On both quilts the #1 triangle is cut from a dark fabric used in the quilt's center. Only three fabrics from the quilt's center are used. Adding this 8" border to a 72" quilt results in an 88" finished quilt.

MATERIALS
• 1 yard fabric for pieces #1 and #6
• 1 3/4 yards fabric for pieces #2, #4 and #5
• 3/4 yard fabric for piece #3

INSTRUCTIONS
1. Piece units as shown in Figure 1. Repeat for 17 units

to complete one side strip. Complete four strips.

2. Piece four corner units referring to Figure 2.

3. Sew a strip to each side of the quilt. Using #1 triangles, sew corner units on to complete border.

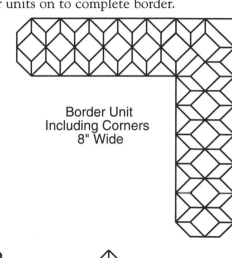

Border Unit
Including Corners
8" Wide

Add seam allowance to all pieces.

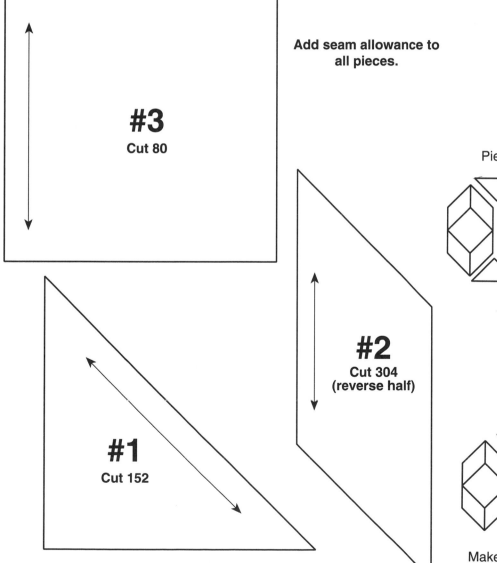

#3
Cut 80

#2
Cut 304
(reverse half)

#1
Cut 152

Figure 1
Piece 1 unit as shown.

Figure 2
Join with #1 triangles.

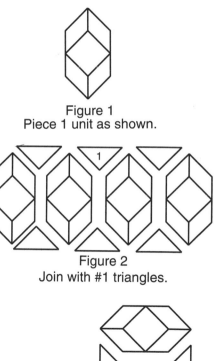

Figure 3
Make corners as shown. Join with side units using piece #1

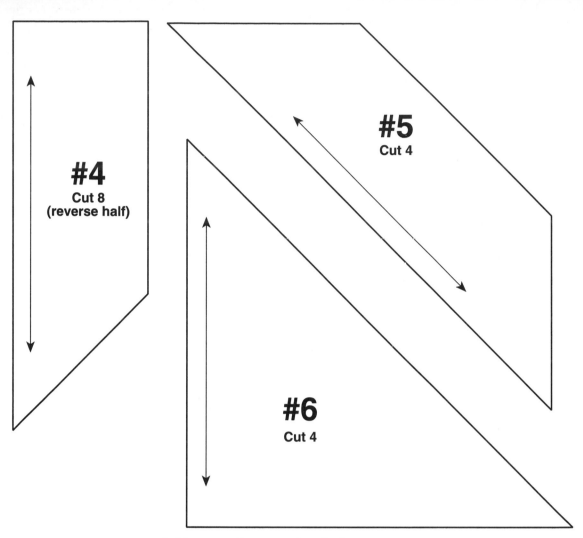

#4
Cut 8
(reverse half)

#5
Cut 4

#6
Cut 4

Add seam allowance to all pieces.

BORDER 2

This border design was used on the Royal Stars of Virginia and Oregon. It is an easy border to piece because it has only a few pieces. Use this 7 1/2" border design for any 75" quilt center to make a 90" finished quilt.

MATERIALS
- 1/4 yard fabric for piece #5
- 1/4 yard fabric for pieces #2, #3 and #6
- 1/4 yard fabric for piece #4
- 2 1/2 yards fabric for piece #1

INSTRUCTIONS
1. Join five #1 pieces with #2.
2. Sew a #3 piece to each end of the pieced unit as shown in Figure 1. Repeat for four side border strips.

3. Sew #4 to #6. Repeat for the reverse pieces. Join together with piece #5. Repeat for four units for corners referring to the Figure 2.

4. Sew one long border strip to two opposite sides of a 75" quilt top. Sew a corner square to each end of the remaining two strips; sew to the top and bottom of the quilt top to finish.

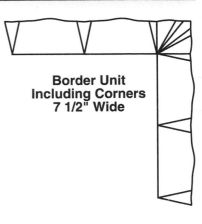

**Border Unit
Including Corners
7 1/2" Wide**

Figure 1
Sew 5 #1 pieces together with #2; add #3 on each end.

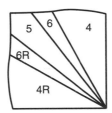

Figure 2
Piece 1 corner unit as shown.

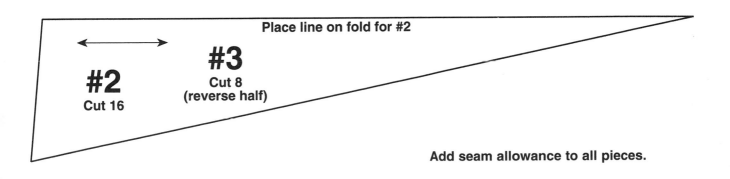

Place line on fold for #2

#2
Cut 16

#3
Cut 8
(reverse half)

Add seam allowance to all pieces.

Add seam allowance to all pieces.

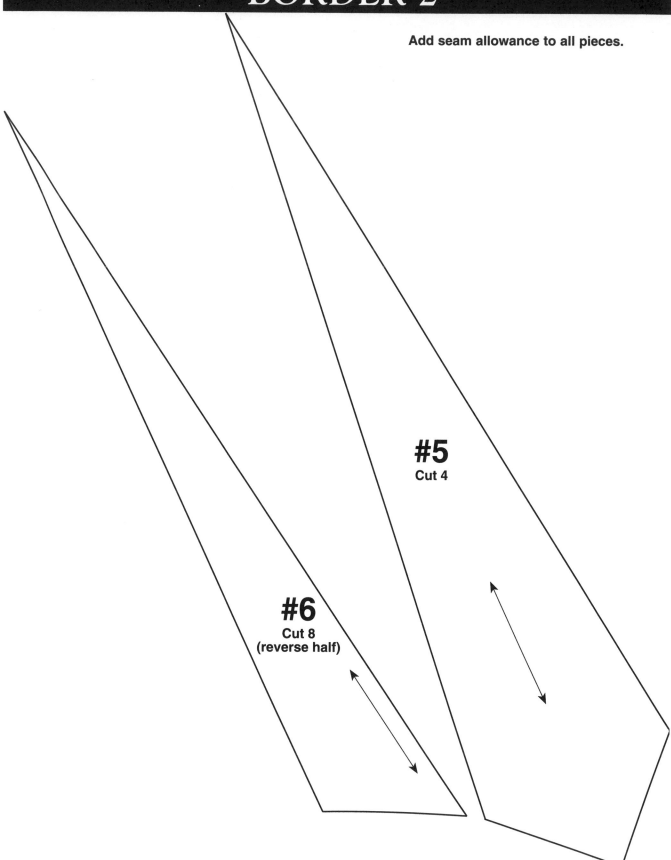

#5
Cut 4

#6
Cut 8
(reverse half)

Add seam allowance to all pieces.

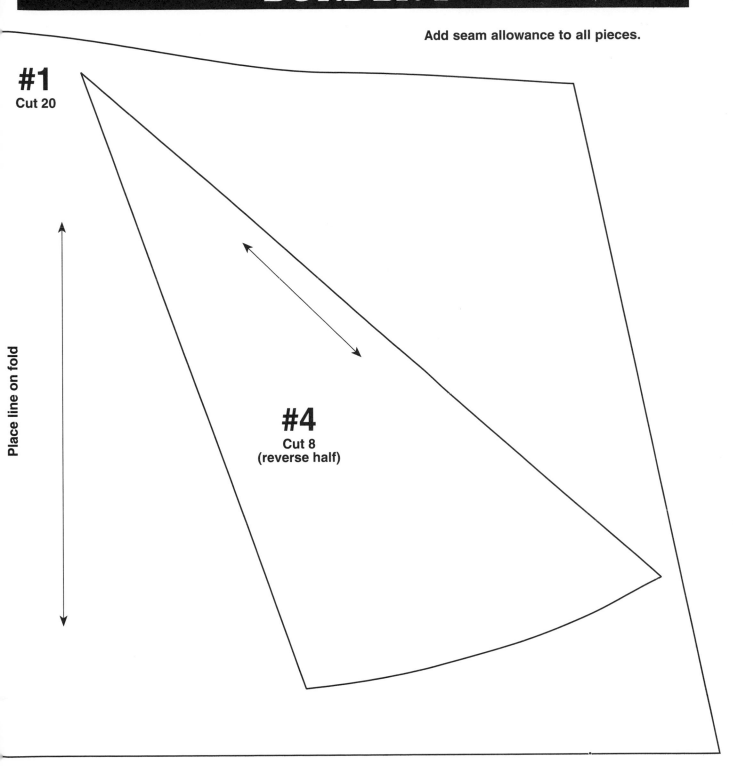

#1
Cut 20

Place line on fold

#4
Cut 8
(reverse half)

This border design was used on the Royal Stars of Pennsylvania, Utah, Missouri, Vermont and Colorado. Each quilt except Pennsylvania is a 72" square; it is 75" square. To make this border fit a 75" center, increase the number of units per side by one and add 1" (1/2" to each side) borders to make the center 76" square. Adding this 5 3/8" border to a 72" center results in an 82 3/4" square center. Adding it to a 78" center results in an 88 3/4" square center.

MATERIALS

- 1 yard fabric for pieces #1 and #3
- 1 1/4 yards each 2 fabrics for piece #2

INSTRUCTIONS

1. Join two #2 diamonds; set in a #1 triangle as shown in Figure 1. Repeat for 19 units.

2. Join the 19 units; set in remaining #1 triangles as shown in Figure 2. Repeat for four border strips.

3. Sew a strip to each side of the quilt center.

4. Sew #2 pieces together at the corners and set in #3 squares as shown in Figure 3 to complete quilt.

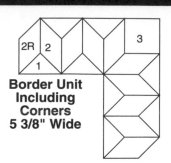

Border Unit Including Corners 5 3/8" Wide

Figure 1
Sew 2 #2 diamonds together; set in #1 triangle as shown.

Figure 2
Set #1 triangles between points of #2 diamonds as shown.

Figure 3
Set #3 squares in at the corners as shown.

Add seam allowance to all pieces.

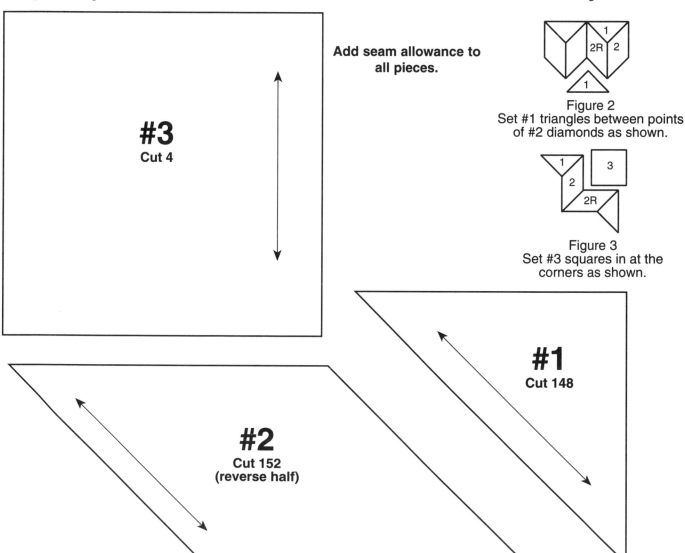

#3
Cut 4

#1
Cut 148

#2
Cut 152
(reverse half)

This border design was used on the Royal Star of Illinois. This border can be used on any 75" quilt top. Adding this border on a 75" quilt top results in a 90 1/2" finished quilt.

MATERIALS

- 1/4 yard fabric for piece #1
- 1 yard fabric for piece #2
- 1 1/4 yards fabric for pieces #3 and #4

INSTRUCTIONS

1. Sew #1 to #2 to #1 as shown in Figure 1; repeat for 17 units.

2. Join the triangle units with piece #3 as shown in Figure 2. Continue sewing units together to finish one strip.

3. Repeat to complete four border strips.

4. Sew a strip to each side, setting in piece #4 at corners.

**Border Unit
Including Corners
7 3/4" Wide**

Figure 1
Sew #1 to #2 to
#1 as shown.

Figure 2
Join the #1–#2 units
with #3 as shown.

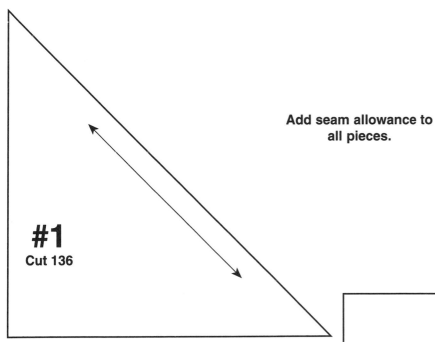

#1
Cut 136

**Add seam allowance to
all pieces.**

#2
Cut 68

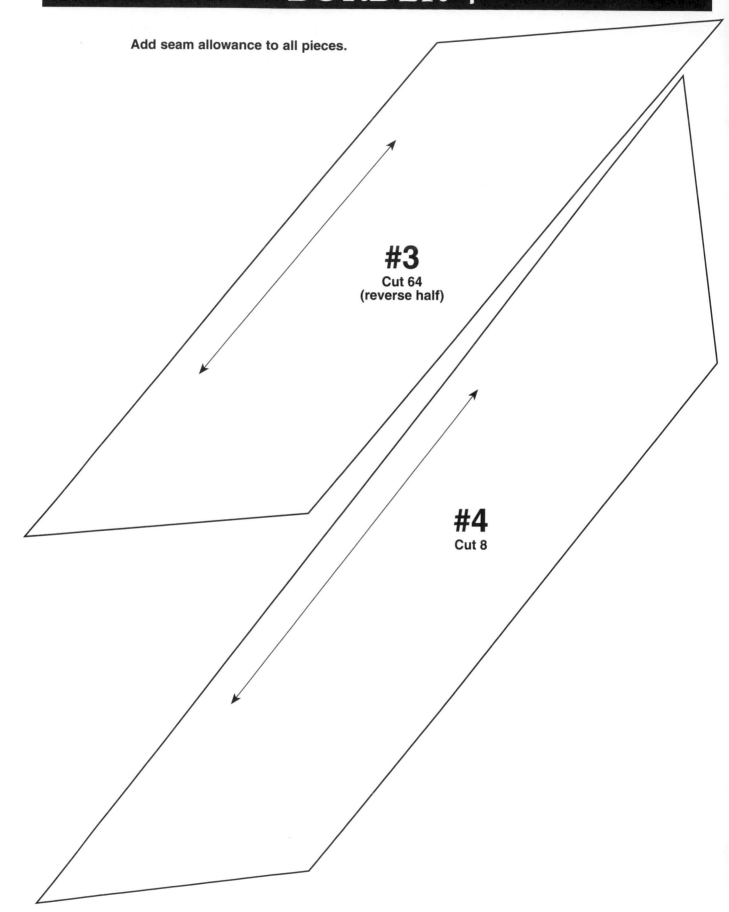

BORDER 4

Add seam allowance to all pieces.

#3
Cut 64
(reverse half)

#4
Cut 8

This border design was used on the Royal Stars of North Dakota, Minnesota and Wisconsin which are all 75" designs. The border was designed to fit a 72" quilt. To make it fit a 75" quilt, enlarge the quilt top to 80" by adding 2 1/2" (cut 3") plain border strips to each side. Add one more unit to the border strip to make it fit the 80" top. This produces a 90" quilt top. Use the design as given to fit any 72" quilt top to finish at 82" square.

MATERIALS
- 1/2 yard fabric for piece #1
- 1/2 yard fabric for pieces #1, #4 and #7
- 3/4 yard fabric for pieces #2 and #5
- 1/2 yard fabric for pieces #3 and #6

INSTRUCTIONS
1. Sew #2 to #3; repeat for reverse pieces. Repeat for 18 units. Join with pieces #1 and #4 referring to Figure 1 to make one border strip. Repeat for four strips.

2. Sew #5 to #6 and #7 to #7 reversed. Sew the pieces together to make a corner unit as shown in Figure 2; repeat for four units.

3. Sew a long strip to the two opposite sides of the pieced quilt center. Sew a corner square to each end of the remaining two strips. Sew to the remaining sides to complete the quilt top.

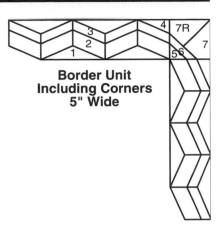

Border Unit Including Corners 5" Wide

Add seam allowance to all pieces.

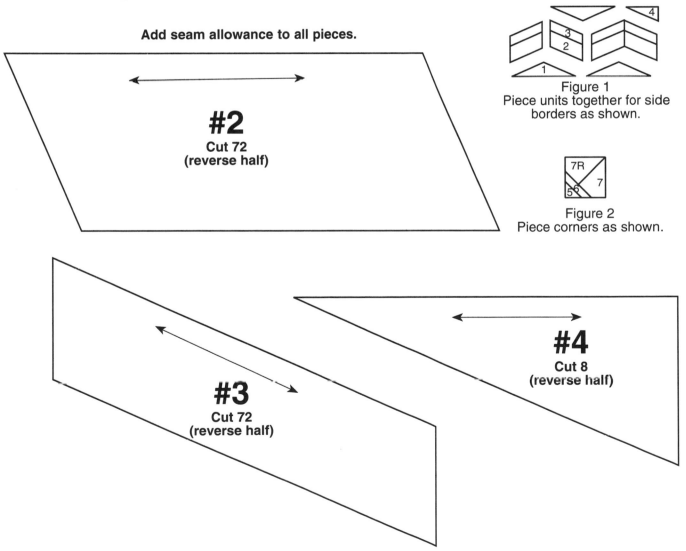

#2
Cut 72
(reverse half)

Figure 1
Piece units together for side borders as shown.

Figure 2
Piece corners as shown.

#3
Cut 72
(reverse half)

#4
Cut 8
(reverse half)

Add seam allowance to all pieces.

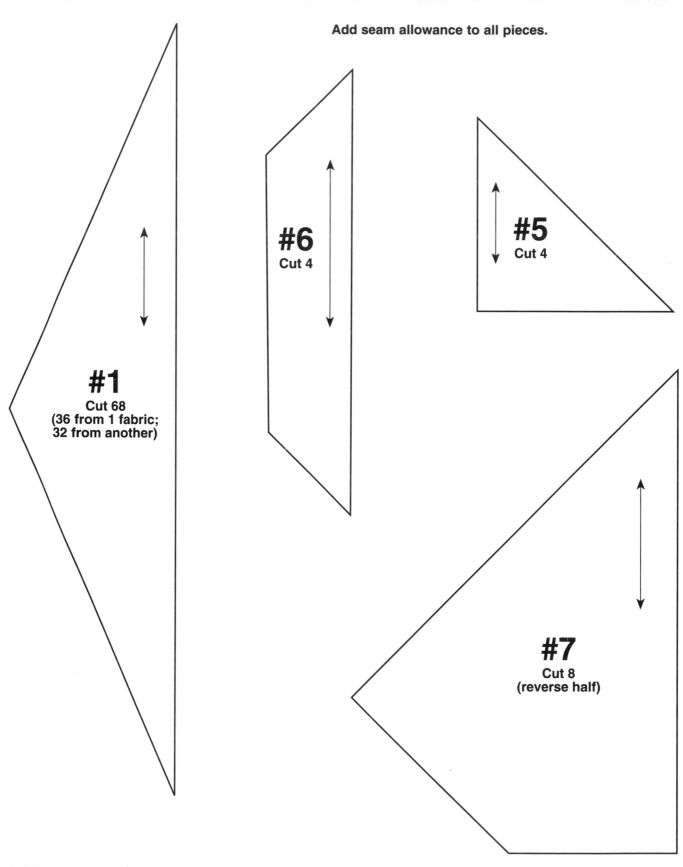

#1
Cut 68
(36 from 1 fabric;
32 from another)

#6
Cut 4

#5
Cut 4

#7
Cut 8
(reverse half)